Seshat

VOLUME 9

Seshat

VOLUME 9

Contributors:
Ishpreet Chana
Peter Anto Johnson
John Christy Johnson
Minahil Syed
Benjamin A. Turner
Paawan Virdi.

Editors:
Daivat Bhavsar
Austin A. Mardon
Catherine Mardon
Susie Woo

Cover Design:
Susie Woo

First Printing: 2021

Cover Design by Susie Woo
Typeset by Josh Harnack

ISBN: 978-1-77369-658-4
Ebook ISBN: 978-1-77369-659-1

Golden Meteorite Press
103 11919 82 St NW
Edmonton, AB T5B 2W3
www.goldenmeteoritepress.com

GM★
P R E S S

Table of Contents

Benefits and Limitations of Neural Network Theories

Ishpreet Chana

Department of Chemistry and Chemical Biology, McMaster University

Introduction

The human brain is a complex organ that is part of the central nervous system. It functions as a control center for releasing hormones, storing information, regulating blood pressure, etc. The brain is utilized as a model to synthesize Artificial Neurons which are used to generate various Neural Network systems. Neural networks include: Feedforward Neural Networks, Radial Basis Function Neural Network, Recurrent Neural Network, and Convolutional Neural Networks. These computational systems have had a significant impact in various applications, such as Google translate. Hence, this article will discuss the benefits and limitations of each specific Neural Network.

The neuron (sensory, effector, and interneurons-types) is composed of complex structures which work together to stimulate our bodies. These neurons are typically composed of dendrites, a cell body (soma), and an axon. These structures facilitate together to transmit electrochemical information through the change in the membrane potentials which lead to synapses through the presynaptic and postsynaptic neurons. The cells further receive information and transfer the outputs of that neuron as the inputs for the next neuron. The Artificial Neural Networks (ANN's) are mathematical based models that are derived upon the structure and functions of these biological neurons. This allows ANN's to stimulate the way the brain processes information and to mimic the behaviour of an actual neuron. From ANN's, the Neural Network Theories (NNT's) are created and utilized to further understand how the neurons within the brain and central nervous system work to generate artificial intelligence in technology. These neural networks are generally composed of thousands of artificial neurons that are connected by nodes and contain three significant layers: one input layer, one or many hidden layers, and one output layer. The inputs are referred to as (Xn) and are multiplied by a weight value (Wn), where the sum of the input and weight is passed through the activation function. The activation function allows the input to be converted to the output, which is referred to as the activation value. However, to determine the appropriate inputs, a pre-experiment estimation step is required. This step is called a "Forward pass" step, in which prior to the network predicting the output, the training phase is initiated, where the weights are paired with the artificial neurons. Then the inputs are transported into the network and the activation of all the nodes in the hidden layers are established to further determine the output value. Additionally, since the weights are randomly assigned, a background propagation step is also conducted to determine the associated error. The error is calculated using the cost function, which adjusts the weights by going backwards from the outputs to the inputs to obtain the possible outcome which produces the lowest amount of error.

FeedForward Neural Network

The Feedforward Neural Network (FNN) was the first discovered neural network that contained the simplest design. Within this particular type of artificial neural network, the connections that the nodes form do not generate a cycle because each node has a different composition. Therefore, the signal moves through a definite pathway going from the input layer to the output without any loops. Furthermore, there are 2 types of FNN's: the Single-layer perceptron and the Multi-layer perceptron. A perceptron is an algorithm used for the learning of binary classifiers (it can decide whether the input belongs to a specific class or not) and is composed of 4 different parts: the input values, the weights, the net sum, and an activation function. Moreover, the single layer perceptron does not possess any nodes, hence it only contains one layer and is only prone to solving linear functions. Whereas the multi-layer perceptron possesses one or more hidden layers which have many perceptrons and can potentially use both linear and nonlinear functions. Both of these neural networks provide multiple beneficial applications and some limitations. Single layer perceptrons were used in a study where researchers used the neural network to create a method called "Privacy-preserving single-layer perceptron" which used a hormo-morphic paillier cryptosystem. This technique is primarily used for obtaining a disease risk model from a cloud service without leaking essential information in hospitals. Moreover, in an article, Multilayer perceptrons were utilized to identify amyloid proteins. The method created was called PredAmyl-MLP and it was able to achieve an accuracy of 91.6% for identifying the amyloid proteins. In addition, Multi-layer perceptrons can also help in various aspects in architecture and structural engineering through the activation function. For example, the usage of Multi-layer perceptrons, containing two hidden layers can identify damage in bridge structures. Some limitations between these methods are that the Single layer perceptrons can not process complex inputs compared to the Multi-layer perceptrons and for the Multi-layer perceptrons it can be difficult to determine certain parameters due to its complexity in structure. Moreover, the Single-layer perceptron can only be applied to linear functions, while the Multi-layer perceptron can compute both linear and nonlinear functions. This makes the Multi-layer perceptron more useful in certain applications.

Radial Basis Function

The Radial Basis Function Neural Network (RBFNN) is a type of Feedforward Neural Network. It also consists of three layers: an input layer, one hidden layer which has non-linear activation units called the "Radial Basis Function activation function", and an output layer. Just like the Single-layer perceptron, the RBFNN has a simple structure, which makes it easy to use as it lacks complexity compared to Multi-layer perceptrons. The RBFNN has many advantageous applications in a variety of areas. For instance, in a study the RBFNN was used to analyze the power output generated from a wind turbine and to improve the efficiency of the RBFNN, a clustering algorithm called "Particle Swarm Optimization (PSO)" was utilized with adjustments made accordingly towards the parameters. Additionally, it can be used in the medical industry. For example, the RBFNN was used to diagnose diabetes mellitus based on the insulin content. Moreover, there are many limitations to this network. These include being highly sensitive to the dimensionality of the nodes, the classification is slower compared to the multi-layer perceptron since each node in the hidden layer has to process the RBF function, and it also can not compute large complex sequences, due to the simplicity of the network design.

Recurrent Neural Network

Recurrent Neural Networks (RNN) are derived from the Feedforward Neural Networks as well. They form connections between the nodes in the hidden layer which form a cyclic loop, to allow the system to repeat previous outputs to be utilized as the inputs. This characteristic allows them to be able to memorize historical information for long periods of time. These neural networks have significant impacts in technology development. It can be applied towards music generation, name entity recognition (passwords), machine translation, etc. For example, the speech recognition translator has further impacts in developing Google translate, and other websites with translations, such as Amazon. Additionally, in a study, the RNN was used in identifying misspelled words, through using the memory of previous handwriting documents and was found to achieve approximately 84.5% accuracy. Long short-term memory (LSTM) is a type of RNN and it can also be used for the recognition of sequences over a long period of time. However, this particular RNN uses four neural networks and is composed of three gates: the Forget gate, the Input gate, and the Output gate. LSTM has significant applications towards the medical field. For instance, it can be applied in protein homology detection based on building blocks referred to as long short term-memory. In this process, the long short-term memory can identify the relationship and patterns between the protein sequences to further synthesize the protein copies. There are also various limitations that arise within both the RNN and LSTM. Both can be very slow at times, have difficulties processing large sequences to memorize, and it can be very difficult to train the RNN to produce applications.

Convolutional Neural Networks

Convolutional Neural Network (CNN), contains multiple hidden layers like the Multi-layer perceptron. In this network, it is highly selective towards the depth, width, and height of its layers. The top layer of CNN contains: The math layer, the rectified linear unit layer, and a fully connected layer. The Math layer's purpose is to analyze the image pattern and essentially assign a number towards each unit, and then a filter is applied to a corner of the image and eventually the whole image is processed throughout. This causes the data to be arranged in a three-dimensional array and then the Rectified Linear Unit (ReLu) layer is composed of the activation function which further analyzes the image in greater detail. Lastly, the fully connected layer, takes the outputs from the ReLu layer and produces a "N"-dimensional vector output. The "N" signifies the program that is being identified and the program allows for the image to be broken down into smaller pieces and approves it to be analyzed. For instance, if the image is of a cat, the program would analyze for relevant features such as four legs, pointy ears, a tail, etc., and eventually after analyzing all the features, it would generate the entire image. This property allows these networks to be useful in the creation of many new technologies. It is applied to facial recognition, in which many electronics use, such as "Apple" products as it is easier to unlock devices instead of having a password. CNN is predominantly used for analyzing and processing visual imagery. Moreover, this neural network can be applied towards other inputs of data such as predicting the climate. For instance, CNN has been used on climate data sets, in order to predict when the next disaster will likely occur based on the general trends in the patterns and it was approximately 89-99% efficient at detecting natural disasters. There are many limitations that arise in this network based on its structure. One challenge is that the processing of the image is a difficult task because it is highly sensitive and specific. For example, it can not process the same image taken from different angles, lighting conditions, and different backgrounds. Additionally, it can be time consuming, as it takes longer to process more complex images.

Conclusion

In conclusion, the FNN's have 2 different types of neural networks: the Single-layer perceptron, the Multi-layer perceptron, and the FNN is used as a basis to generate the RBFNN, RNN and CNN. The Single-layer Perceptron and the RBFNN only have one hidden layer and the Multi-layer perceptron and CNN have multiple hidden layers. These properties make them each useful towards specific applications. Lastly, the RNN is a cyclic network in which it can store information for long periods of time making it more useful in certain applications compared to the FNN's. However, each of these Neural Networks have their own drawbacks, but each provide a wide range of applications that are useful in everyday life and can be used to create further advancements in the future.

References

(1) Buscema, M. (1998) Theory: Foundations of Artificial Neural Networks. *Substance Use & Misuse 33*, 17–199.

(2) Fernández, S., Graves, A., and Schmidhuber, J. (2007) An Application of Recurrent Neural Networks to Discriminative Keyword Spotting. *Lecture Notes in Computer Science Artificial Neural Networks – ICANN 2007* 220–229.

(3) Li, S., Chen, J., and Liu, B. (2017) Protein remote homology detection based on bidirectional long short-term memory. *BMC Bioinformatics 18*.

(4) Li, Y., Zhang, Z., Teng, Z., and Liu, X. (2020) PredAmyl-MLP: Prediction of Amyloid Proteins Using Multilayer Perceptron. *Computational and Mathematical Methods in Medicine 2020*, 1–12.

(5) Pandey, P., and Barai, S. (1995) Multilayer perceptron in damage detection of bridge structures. *Computers & Structures 54*, 597–608.

(6) (2014) Types of Artificial Neural Network. *Medical Diagnosis Using Artificial Neural Networks* 58–67.

(7) Wang, G., Lu, R., and Huang, C. (2015) PSLP: Privacy-preserving single-layer perceptron learning for e-Healthcare. *2015 10th International Conference on Information, Communications and Signal Processing (ICICS).*

(8) Wang, G., Lu, R., and Huang, C. (2015) PSLP: Privacy-preserving single-layer perceptron learning for e-Healthcare. *2015 10th International Conference on Information, Communications and Signal Processing (ICICS).*

How the Animations Industry has Evolved

Ishpreet Chana

Department of Chemistry and Chemical Biology, McMaster University

Introduction

Animation has come a long way since the nineties and have made many beneficial impacts on society. Whether they cheer an individual up, or make one cry, they help invoke emotions and, in some cases, portray real life scenarios. For instance, the anime movie "A Silent Voice ", showcased a girl who gets bullied due to her impaired hearing disability and outlines the consequences of bullying. Over the years, various animation studios have made multiple advancements in developing better animation quality for shows and films. In fact, the first ever animation development was in 1831, in which the Phenakitstoscope was developed. This was a machine which produced a moving person through the rotation of small disks. The purpose of this article is to emphasize the history of animation and how the animation industry has evolved through the development of various techniques.

Evolution of Cinema

The "Silent Era" is where the cinematic industry first rationalized. Within this Era, the films were shown using the Cinematographe and through reels of tape. Through this technology, the animations were only portrayed in black and white with sound being nonexistent. During the "Silent Era", various discoveries were made in the animation industry. For instance, the Fleischer Brothers established the rotoscope and through it they produced their winning hit "Koko the Clown". The rotoscope was a method, in which animators traced over the footages of the frames to produce animated films. This technique quickly expanded to other studios, such as Disney to produce various films, and in Modern times, rotoscoping is still used to produce animations, although done through computers. This Era helped shape the animation industry through the many advancements of techniques and picture quality. Moreover, the "Silent Era" ended in 1929, giving rise to the "Talkies" Era, once sound was enforced into animations. The first movie with sound associated with this was the "Jazz Singer". This was done using a sound film system developed called the Vitaphone. Furthermore, Walt Disney Studios created a device to produce sounds called the Sonivox. This device was composed of two cylinders, similar to tin cans, which one side was held towards the individual's mouth. This device was set as a model to create the vocoder, which is used to produce Modern day animations. Moving on, in 1932, colour was introduced into movies by the Technicolour technique. This method uses a three-strip camera which portrays the scenes in cyan, yellow, and magenta. This allowed the films to be vibrant in colour, making them appealing and paving the pathway for further improvements in animations.

Technology

The first ever animation method was first discovered by Emile Cohl, in which the animations were all hand drawn and was utilized in the 1908 movie called "Fantasmagorie". This technique utilized black negative film in which stick figure characters were drawn onto approximately 700 panels in "Fantasmagorie" which was then animated. During 1914-1967, animation was typically conducted by hand drawing the characters using the Cel animation technique. In this technique, the Cell is composed of cellulose nitrate and camphor, which causes it to be transparent, making it easier in the animation process. For example, in 1914, "Colonel Heeza Liar" was an animation that utilized this technique. Other techniques, such as keyframing, registration marks, and animation loops were also used in the 19th century as well. Since animations contain many frames, Keyframes contain specific information about the starting and ending of an action, which allow the various frames to be observed as a moving action. Registration marks were used to help stabilize the backgrounds, so they would not appear flashy and animation loops caused the frames to repeat, making it helpful in reducing the frames, to animate longer movies. "Gertie the Dinosaur" was an animated film that used these three techniques, which all set a basis for various other animated films.

Disney

Walt Disney revolutionized the animation industry by producing many advances in animation. In 1937, "Snow White and the seven dwarfs" was the first full length animated film and the first Disney princess movie released. This film was created using the Cel animation technique with creating colour. After each drawing was made to the Cel sheets, the back of the sheets were painted with colour to retain the outline of the drawings in ink. The glass backgrounds in which the sheets were pasted, were painted with watercolour. Additionally, the animation film was the first to utilize the "multiplane" camera technique, which moves the sheets past the camera at relatively high speeds to create depth and complexity in the film. The "multiplane" camera was used for various other Disney productions, such as "The Little Mermaid", "Peter Pan", "Cinderella", and "Alice in Wonderland". Moreover, in Disney's thirty-first film, "Aladdin", used a hybrid of both the hand-drawn animation and computer animation and the software they utilized to create the 3-dimensional animation was Pixar's Renderman. Computer animation is a relatively easier method to create animations. For instance, both 3-dimensional and 2-dimensional animation can be with it and if a mistake is made, it can easily be corrected by erasing, compared to the traditional hand-drawn techniques. In 1995, Disney released "Toy Story", which was the first full length animated movie using Computer Generated Imagery (CGI), and this caused the animation quality to appear to be different compared to the others. The CGI technique is derived from computer animation, but it can process more dynamic and complex images. In the animation process, it primarily uses the rendering technique, which saves the image frame with lighting and colour. To animate "Toy Story", the animators had 117 computers which all ran 24 hours a day to render approximately 114,240 frames using a new software created at the time called Render-man. However, there were some limitations in the animating procedure. For instance, at the time the animators did not comprehend how to animate human clothing within the film, as it was time consuming. But through the sequels such as "Toy Story 2", advancements in the animation had been made, giving rise to more realistic characters. Lastly, in 2010, Disney released "Tangled" which was primarily animated using Computer Generated Imagery (CGI) and non-photorealistic rendering. The software that is used for animating affects the overall quality of the animation, in the CGI technique. In "Tangled", the animators used ZBrush, to create more in-depth scenes with realistic characters. Therefore, films like "Tangled" have a different animation style due to the type of software utilized, despite using the same animation method.

Anime

Anime originated in Japan from its heavy influence based on the American animation industry. The first ever anime known was created in 1952 called "Astro Boy". This was animated using the Cel sheets and placing them onto various backgrounds in order to animate it. Anime began to utilize computer animation in the late nineties, in which the first computer animated anime movie was released called "Golgo 13: The Professional". Moreover, nowadays the more preferred technique in anime production is the computer animation and CGI techniques, to grasp more complexity and depth in the overall quality. Additionally, the CGI method is also impacted by the type of software used to animate. For instance, the first three seasons of "Attack on Titan" was produced by WIT studio and the fourth season was produced by MAPPA. However, the MAPPA production had higher quality in the anime's character design and greater detail in the scenery. Despite the fact, these animation studios both used CGI animation, each used different software's in the animation process which ultimately affected the overall quality of the animation. Some animation studios like Studio Ghibli, still use traditional animation methods like the Cel technique but incorporate CGI for more difficult and detailed scenes.

Genre

Throughout the years, the genre of animations has expanded to greater heights. For instance, Disney princess films back in the nineteenth century typically involved a prince to save the princess, such as "Snow White", "Sleeping Beauty", etc. These films emphasized the gender roles and the subordination of men within their lives in that time period. However, in the twentieth century, Disney evolved to eliminate these portrayed gender stereotypes. This is seen in films, such as "Brave" and "Frozen", where the princess is able to protect herself, without relying on men to save them. Alongside this, Anime has also evolved in incorporating more genres in their animations, such as: slice of life, romance, supernatural, sports, etc. In addition, there have been multiple anime's produced regarding the LGBTQ+ community, such as "Given". Therefore, animations nowadays showcase a variety of different topics and struggles portrayed in real life.

Conclusion

In conclusion, there have been many new technologies created to produce a wide range of animations. This can be seen from the traditional hand drawing methods to the production of computer animation technologies. The development of computer animation has allowed many animation industries to create several unique films and have made it easier to animate longer productions. Additionally, it was set as a basis to develop the CGI technique, allowing more complexity and detail to be added to various animations. Furthermore, Walt Disney Studios and Anime studios in Japan have both evolved over the years, in animation techniques and the genres of animation have risen to greater heights. Overall, animation has had tremendous improvements throughout the years, and will continue to improve high quality animations in the years to come.

References

Anime is Born and Invades America (and American TV goes to Japan and Influences Manga/Anime). (2016). *Manga and Anime Go to Hollywood.* doi:10.5040/9781501312755.ch-006

Avila, E. (2018). American Cultural History: A Very Short Introduction. *Very Short Introductions.* doi:10.1093/actrade/9780190200589.001.0001

Batchelder, D. (2016). Snow White and the Seven Dwarfs: *Master Score by Walt Disney.* Notes, 73(1), 157-161. doi:10.1353/not.2016.0106

Brown, N. (2018). Toy Story and the Hollywood Family Film. *Toy Story.* doi:10.5040/9781501324949.ch-002

Corsaro, S. M. (2002). The evolution of animation. *ACM SIGGRAPH 2002 Conference Abstracts and Applications on - SIGGRAPH 02.* doi:10.1145/1242073.1242332

Dundes, L. (2020). The Upshot on Princess Merida in Disney/Pixar's Brave: Why the Tomboy Trajectory Is Off Target. *Humanities, 9*(3), 83. doi:10.3390/h9030083

Finan, K. (2017, January 08). The History of Animation Sound. Retrieved from https://www.boomboxpost.com/blog/2015/11/8/the-history-of-animation-sound

MasterClass. (2020, December 01). History of Animation: How Animation Evolved Over a Century. Retrieved from https://www.masterclass.com/articles/a-guide-to-the-history-of-animation

The Prevalence of Hate Crimes

Ishpreet Chana

Department of Chemistry and Chemical Biology, McMaster University

Introduction

Over the years, hate crimes have risen to greater heights, impacting many lives, such as causing individuals to live in fear. A hate crime is a criminal act which is targeted against social groups and is perceived as a prejudice or bias motivated crime. Social groups can include religion, sexual orientation, disabilities, language, etc. According to the FBI in 2017 within America, there were 7,145 hate crimes reported. From this, 58.1% of the hate crimes were regarding race, 16% were regarding sexual orientation, 22% regarding religion, 0.6% regarding gender identity, and for disabilities it was 1.6%13. Despite the fact of certain laws being put in place, hate crimes still exist. The purpose of this article is to emphasize the hate crimes targeted against the LGBTQ+ community, Race and Religion, and Disabilities and how they have impacted society.

LGBTQ+ Hate Crimes

In Canada, the Charter of Rights and Freedom protects individuals from experiencing discrimination based on race, religion, sexual orientation, etc. Despite the Laws being put into place, individuals still experience hate crimes around the world. For instance, the legislation for equality was passed in 1982 in China. Despite the law coming into place, hate crimes still prevail in China. In the nineteenth century, Homosexuality was classified as a mental disorder in the Chinese Classification of Mental Disorders (CCMD), but was removed in 2001. However, in some parts of China, Homosexuality is still considered a mental disorder, where LGBTQ+ individuals are forced to undergo treatments. Additionally, some companies in China do not provide benefits to LGBTQ employees. In one case, in 1998 within the United States of America (USA), an anti-gay hate crime occurred against a student named Matthew Shepard, in which he was brutally murdered by two individuals. Despite the tragic death of Matthew Shepard, this led to great advances in gaining LGBTQ+ rights. For instance, Matthew's story was shared to thousands of people through a play called "The Laramie Project" and Matthew's family have created a foundation to help fund programs regarding the discussion of sexual orientations14. Moreover, in a study, it was found that Trans individuals are more prone towards violence, threats, and anxiety compared to non-trans individuals. It was noted that approximately 56% of transgender individuals feel unsafe in public and tend to cover up their gender identity to avoid being victimized against a hate crime. Moreover, Transgender people are twice as more likely to think about suicide and have a higher suicide rate than the LBG community9. This is due to the various challenges that arise in their lives such as: isolation from society, experiencing transphobia, inequality, etc. However, certain protective measures have been put in place to help transgender individuals feel safer. These include enforcing clubs in school such as the "Gay-Straight Alliance" (GSA). For example, in Alberta they passed a Bill-24 which is a legislation that allows students to form the GSA clubs without school interference15. Additionally, many places have bathrooms and housing specifically designed for Transgender individuals.

Race and Religion

George Floyd was a black man, who was arrested by a white police officer on May 25th, 2020 and during the arrest the officer was kneeling over his neck which led to his death. The George Floyd case left many people with rage leading to a series of riots and protests in the USA. The Black Lives Matter movement has spread awareness about the discrimination against black individuals and this led to various protests being conducted around the world. Moreover, Anti-Semitic (prejudice against the Jewish community) hate crimes have increased over the years and in 2019, the FBI in the USA stated that there was a 14% increase from 2017-20183. In 2018, there was an incident regarding Anti-Semitism, where a gunman entered the Tree of Life Synagogue and killed 11 individuals3. This was known to be one the deadliest attacks targeted against Jews. However, over the years some countries have implanted certain rules to help fight Anti-Semitism. For example, Austria has presented new enforcements to prevent Antisemitic hate crimes from occurring1. This includes providing education about Judaism and implementing stricter prosecution against Antisemitic hate crimes. Furthermore, due to the Coronavirus outbreak, Asian-Hate crimes have also increased. For instance, physical assaults against both young and old Asian Canadians have been reported, in which they were spat or coughed on, or rocks were thrown at them. In America, an 89-year-old Chinese woman was brutally slapped and set on fire by two individuals in Brooklynn, New York. Moreover, In California, a Filipino woman who was a medical care worker was shoved to the ground by another individual who explicitly told her to go back to China5. However, many countries around the world are implanting on rules to try and help Asians feel safer. For example, on April 22nd, 2021, the Senate in Washington AP, passed a legislation to eliminate Asian-Hate crimes against Asian Americans with implanting stricter prosecutions5. In addition, education programs are being developed to help gain awareness against the situation. Lastly, Islamophobia has also increased in some parts of the world, and it refers to prejudice towards Muslim people. In an article, a black Muslim woman was attacked by another individual in Alberta. In this incident the woman was hit on the head with a shopping bag, while receiving racial slurs16. To eliminate Islamophobia, there have been various actions put into place, such as educational classes, various protests, etc. Additionally, the "National Council of Canadian Muslims" published a book regarding how to fight Islamophobia and its potential effects on children2.

Disabilities

Hate crimes against individuals with developmental disabilities are reported to be higher than individuals without disabilities. They experience a variety of hate crimes such as: assaults, verbal abuses, property damages, sexual harassment, etc. In one case, an individual named Daniel Smith had Autism and the police did not take him seriously after he reported the disability hate crime, in which he was insulted, in Exeter and Northamptonshire. Instead, the police locked him up in a cell for approximately eight hours. However, after that incident, he was brutally attacked by two men, who punched and kicked him, and left him with bruises, as well as a dislocated shoulder6. Afterwards, Daniel attempted suicide due to the traumatic incident which left him depressed and concerned about his safety. It took the police one month to declare the incident as a disability hate crime. This emphasizes that people with disabilities struggle to gain justice and have their cases disregarded. However, some organizations have been developed that help individuals who have experienced a disability hate crime including the Equality and Human Rights Commission, the VictimLink BC, etc.

Conclusion

In conclusion, the LGBTQ+ community, individuals with different races and religions, and people with developmental disabilities all face various hate crimes every day, which impacts their lives in negative ways. Despite there being certain legislation enforced in various countries, hate crimes still prevail against the LGBTQ+ community, race and religion, and individuals with disabilities. However, little by little, certain movements have made great progress in spreading awareness regarding the topic of hate crimes and safe spaces have been created to help individuals feel more comfortable to get past their fear of being targeted.

References

(1) (n.d.). Retrieved from https://abcnews.go.com/International/wireStory/austria-presents-national-strategy-anti-semitism-75401052

(2) 5 things your congregation can do to stop Islamophobia. (2021, March 04). Retrieved from https://www.afsc.org/5-things-your-congregation-can-do-to-stop-islamophobia

(3) Antisemitism and Hate Crimes. (2020, December 16). Retrieved from https://rac.org/issues/antisemitism-and-hate-crimes

(4) Byne, W. (2015). LGBT Health Equity: Steps Toward Progress and Challenges Ahead. LGBT Health, 2(3), 193-195. doi:10.1089/lgbt.2015.0084

(5) Covid 'hate crimes' against Asian Americans on rise. (2021, April 22). Retrieved from https://www.bbc.com/news/world-us-canada-56218684

(6) Crime Victims with Developmental Disabilities. (2001). doi:10.17226/10042

(7) GLAAD Media Reference Guide - In Focus: Hate Crimes. (2016, October 26). Retrieved from https://www.glaad.org/reference/hatecrimes

(8) Jenness, V., Levin, J., & Mcdevitt, J. (1994). Hate Crimes: The Rising Tide of Bigotry and Bloodshed. Contemporary Sociology, 23(4), 576. doi:10.2307/2076413

(9) Maguen, S., & Shipherd, J. C. (2010). Suicide risk among transgender individuals. Psychology and Sexuality, 1(1), 34-43. doi:10.1080/19419891003634430

(10) Perry, B. (2001). In the name of hate: Understanding hate crimes. New York: Routledge.

(11) Rand, M. R., & Harrell, E. (2009). Crime Against People with Disabilities, 2007: National Crime Victimization Survey. PsycEXTRA Dataset. doi:10.1037/e513982010-001

(12) Rodriguez, J. (2021, March 23). New report details 'disturbing rise' in anti-Asian hate crimes in Canada. Retrieved from https://www.ctvnews.ca/health/coronavirus/new-report-details-disturbing-rise-in-anti-asian-hate-crimes-in-canada-1.5358955

(13) The Psychology of Hate Crimes. (n.d.). Retrieved from https://www.apa.org/advocacy/interpersonal-violence/hate-crimes

(14) The truth behind America's most famous gay-hate murder. (2014, October 26). Retrieved from https://www.theguardian.com/world/2014/oct/26/the-truth-behind-americas-most-famous-gay-hate-murder-matthew-shepard

(15) Transgender people and suicide. (2020, November 26). Retrieved from https://www.suicideinfo.ca/resource/transgender-people-suicide/

(16) Yourex-West, H. (2021, March 26). Why are Alberta's Black, Muslim women being attacked? Retrieved from https://globalnews.ca/news/7721850/hate-crime-alberta-attacks-black-muslim-women/

Comparisons of COVID-19 Vaccines

Ishpreet Chana

Department of Chemistry and Chemical Biology, McMaster University

Introduction

Coronavirus has drastically changed the world, impacting each individual's life. The virus has forced the world to advance its defense against viruses. This starts with understanding the virus then expanding the select arsenal to better fight it. SARS-CoV-2 insights on the structure of the spike protein have aided worldwide during the past year, to develop various vaccines. These vaccines include Pfizer, Moderna, Johnson and Johnson (J&J), AstraZeneca, and Novavax. The SARS-CoV-2 spike proteins are composed of sharp crown-like bumps that extend outward, called spikes. These spikes contain glycoproteins which determine how the virus infiltrates the host cell. The Pfizer and Moderna vaccines both use newer technology, through messenger RNA (mRNA). J&J and AstraZeneca are a carrier-based vaccine that uses a non–activated common flu virus. Lastly, the Novavax vaccine is still under clinical review and is a protein-based vaccine which uses saponin. This article provides helpful comparisons regarding their functions, composition of ingredients, symptoms, efficiencies, and benefits and limitations between all of the COVID-19 vaccines.

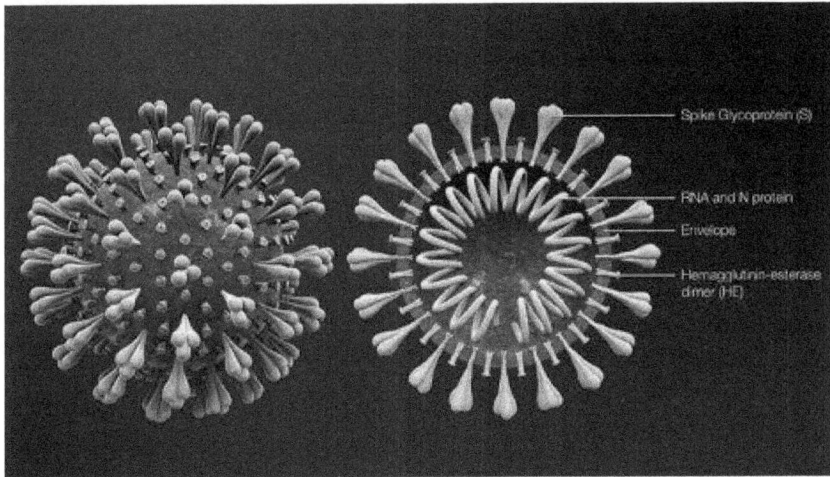

Diagram depicting the structure of the SARS-CoV-2 spike protein.
https://www.biovendor.com/sars-cov-2-2019-ncov-proteins?utm_source=google&utm_medium=organic

How the vaccines work and what they contain

Each vaccine has a unique approach for generating spike proteins. Pfizer and Moderna are mRNA-based vaccines, which contain genetic information for synthesizing the SARS-CoV-2 spike protein. mRNA plays an essential part in synthesizing the proteins in both Pfizer and Moderna in the transcription and translation process. The mRNA carries the genetic coding sequences towards the ribosome and the transfer RNA (tRNA) complementary joins the amino acids to the respective codons. Unlike the Pfizer and Moderna vaccines, the J&J and AstraZeneca vaccines each utilize a different viral vector component to synthesize the spike proteins. The viral vector contains genetic information to produce the spike proteins, using double stranded DNA. J&J has the Adenovirus 26 as the viral vector, which is essentially a virus that causes common flu-like symptoms. Whereas AstraZeneca has a modified version of the chimpanzee Adenovirus (ChAdOxl). Lastly, Novavax is a protein-based vaccine which contains an immune system booster, called saponin. Unlike the other vaccines, a modified spike gene was synthesized and inserted into the baculovirus, which is a DNA virus that infiltrates insects. This infected moth cells and produced spike proteins using the mRNA and tRNA. They were then harvested and transmitted into nanoparticles, which caused them to mimic the structure of the Coronavirus. Furthermore, all the vaccines function in the same manner to generate antibodies to provide protection against the virus. Once the vaccine is injected into the subject's arm, the particles within the vaccine collide between the cells in the body, to produce the spike proteins. In addition, some of the spike proteins form spike protein fragments that migrate towards the surface of the cell. When the vaccinated cell dies, it produces more of the spike proteins and fragments, which are further recognized by the immune system. Antigen-presenting cells (immune cells) absorb the molecules and T- lymphocytes (T cells) and further detect the presence of these spike protein fragments on the surface and signal other immune cells to aid in fighting the virus. In order to generate the antibodies, B lymphocytes (B cells) collide into the spikes on the surface of the vaccinated cells, which allows them to grip onto the spike proteins. The B cells are further activated by the T cells, and they begin to proliferate and generate antibodies that specifically attack the SARS-CoV-2 spike protein. Furthermore, the antigen-presenting cells can also turn on the killer T cells that eliminate the cells that are affected. Moving on, all the vaccines are composed of a medical ingredient and a non-medical ingredient, which have a significant role in an individual's immune system. The major difference between the vaccines is the medical ingredient used. Pfizer and Moderna contain mRNA, J&J and AstraZeneca contain different strains of the Adenovirus, and Novavax contains protein subunits and saponin. However, the medical ingredients are commonly composed of multiple lipids, salts, sucrose, which produce the spike proteins and antibodies, and water which is used for the injection. Each of these ingredients affect how the interactions between immune cells work. For instance, saponin contained in the Novavax vaccine, allows it to generate antibodies, even if the individual has a weak immune system.

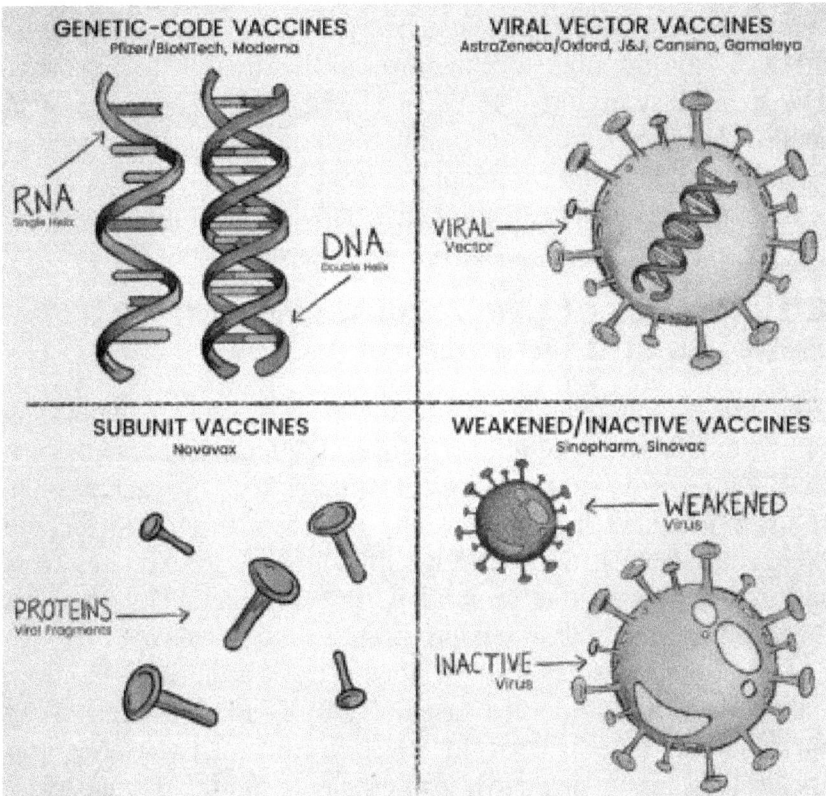

Graphical representation regarding what each COVID-19 vaccine utilizes to produce the spike proteins.

https://abcnews.go.com/Health/covid-19-vaccines-work/story?id=73796121

Doses and Symptoms

The amount of dosage for each vaccine varies depending on its composition and overall effectiveness. Pfizer, Moderna, AstraZeneca, Novavax all require two doses of their vaccines which are spread out at different times. Moderna and Novavax are given in equal 0.5 mL doses, but Pfizer only uses 0.3 mL. The second dose of Pfizer and Novavax is given after 21 days and Moderna is given after 28 days. Moreover, AstraZeneca uses 2 different quantized doses: first dose is 0.2 mL and the second dose is 0.5 mL and it is given 12 weeks after, to allow it to become more impactful. Unlike the other vaccines, J&J only requires a single 0.5 mL dose to provide protection. Consequently, the symptoms after each shot of the vaccines are relatively similar. The common symptoms associated are headaches, pain in the muscles, fatigue, feverish, and possible swelling around the area of the injection. These mild symptoms relatively lessen with proper rest and medication if necessary. According to studies, the mRNA vaccines have been able to trigger anaphylaxis, an allergic reaction, which can be treated with medication such as Epinephrine[17].

Efficiencies of the Vaccines/Variants/ Benefits and Limitations

The efficiency associated with each vaccine fluctuates depending on various factors that affect the overall effectiveness. One such limitation is age, that decreases the impact of all the vaccines due to an individual's immune system. For instance, the Pfizer and Moderna vaccine is only entirely sufficient for approximately ages 18-75 years old[5]. Studies found that the Moderna vaccine was only 86.2% effective after the second dose for individuals over the age of 65[7]. Nevertheless, all the vaccines protect all individuals from hospitalization and severe cases of COVID-19. According to an investigation, the Pfizer vaccine was 85% efficient, two weeks after the first dosage and it led to be 95% efficient after the second dose[5]. Moving on, Moderna was 80.2% efficient within 2 weeks after the first dose and 94% after the second dose[7]. Moreover, J&J was proven to be 77% effective after 2 weeks, but after approximately 3 weeks, it became 85% more effective[14]. AstraZeneca was 62% efficient after the first dose and after the second dose it was stated to possess around 70-80% efficiency if the shot was taken within six weeks[2]. Alongside this, in Canada the "National Advisory Committee on Immunization" indicated that the vaccine should not be used on individuals 55 years old and over[16]. This was speculated due to the formation of rare blood clots reported in a study with 30,000 individuals in Canada. Lastly, Novavax led towards being 96.4% overall efficient in the original strain of the virus[7]. Furthermore, different mutants of COVID-19 have been detected worldwide. Such examples include: (B.1.1.7) which originated from the United Kingdom, (B.1.351) which was recognized in South Africa, and (P.1) which emerged from Brazil. Despite the new mutations of SARS-CoV-2, scientists have been modifying each of the vaccines and conducting clinical trials to eliminate the spread of the variants.

For instance, during a trial the Novavax vaccine was proven to be 86.3% efficient towards the (B.1.1.7), but it was approximately 50% efficient towards the (B.1.351)7,11. Moreover, all the COVID-19 vaccines possess their own potential benefits and limitations. Some benefits of Pfizer and Moderna are that the mRNA within the vaccine is non-infectious and can not cause any harmful impacts. One unique characteristic about mRNA is that it can quickly be synthesized due to its genetic sequencing process. This impacts the manufacturing, as it does not take much to produce one dose. The downsides are that the technology of the mRNA has never been utilized on humans before and requires extensive testing and critical analysis towards the symptoms of the vaccine. Furthermore, the global distribution of the vaccine is a highly sensitive process relative to the vaccines since it has to be stored at drastically low temperatures (-94 degrees Celsius). Moreover, Johnson and Johnson and AstraZeneca also have various advantages. For instance, they possess high specificity of the antigen which allows the vaccinated cells to generate an effective immune response. Additionally, it also only requires one dose, which aids in the storage and distribution of the vaccine and also provides long-term protection overtime. However, the limitation is that individuals may already possess immunity towards the Adenovirus. This can cause a decrease towards the effectiveness of the vaccine, which will stimulate a weaker immune response. Lastly, Novavax contains no live components, composed in the vaccine and it can also be utilized towards individuals that possess weak immune systems. One drawback is that it requires a lot of materials, which leads to problems of shortages and delays in manufacturing.

Conclusion

In conclusion, the vaccines use different approaches to synthesize the SARS-CoV-2 spike protein but function in the same way to produce antibodies. The Pfizer and Moderna vaccines are mRNA based, the J&J and AstraZeneca vaccines are viral vector vaccines, and the Novavax vaccine is protein based. Furthermore, the composition of the vaccines impacts the individual based on how strong or weak their immune system is and the symptoms after the dose are relatively similar. Alongside this, the efficiency of the vaccines is impacted upon age, as older populations have relatively weaker immune systems. Variants have also been identified and potential vaccines are under clinical investigations. Lastly, each vaccine has its potential advantages and disadvantages. Despite the drawbacks of each vaccine, every individual should get vaccinated as it prevents further spread of the virus, allowing everyday life to slowly become more functional.

References

(1) Chambers, C. (2021, April 06). Comparing vaccines: Efficacy, safety and side effects. Retrieved from https://healthydebate.ca/2021/03/topic/comparing-vaccines/

(2) COVID-19 Vaccine Information Sheet. (n.d.). Retrieved from https://health.gov.on.ca/en/pro/programs/publichealth/coronavirus/docs/vaccine/COVID-19_vaccine_info_sheet.pdf

(3) How the Novavax Covid-19 Vaccine Works - The New York Times. (n.d.). Retrieved April 10, 2021, from https://www.nytimes.com/interactive/2020/health/novavax-covid-19-vaccine.html

(4) Iwasaki, A., & Omer, S. B. (2020). Why and How Vaccines Work. Cell, 183(2), 290-295. doi: 10.1016/j.cell.2020.09.040

(5) John-Waller. (2021, February 03). Comparing the COVID-19 vaccines developed by Pfizer, Moderna, and Johnson & Johnson. Retrieved from https://www.boston.com/news/coronavirus/2021/02/02/comparing-the-covid-19-vaccines-developed-by-pfizer-moderna-and-johnson-johnson

(6) Johnson & Johnson's Janssen COVID-19 Vaccine Overview and Safety. (n.d.). Retrieved from https://www.cdc.gov/coronavirus/2019-ncov/vaccines/different-vaccines/janssen.html

(7) Katella, K. (2021, April 07). Comparing the COVID-19 Vaccines: How Are They Different? Retrieved from https://www.yalemedicine.org/news/covid-19-vaccine-comparison

(8) Kaur, S. P., & Gupta, V. (2020). COVID-19 Vaccine: A comprehensive status report. Virus Research, 288, 198114. doi:10.1016/j.virusres.2020.198114

(9) Li, Y., Tenchov, R., Smoot, J., Liu, C., Watkins, S., & Zhou, Q. (2021). A Comprehensive Review of the Global Efforts on COVID-19 Vaccine Development. ACS Central Science. doi:10.1021/acscentsci.1c00120

(10) Mahase, E. (2021). Covid-19: Where are we on vaccines and variants? Bmj. doi:10.1136/bmj.n597

(11) Novavax Confirms High Levels of Efficacy Against Original and Variant COVID-19 Strains in United Kingdom and South Africa Trials. (n.d.). Retrieved from https://ir.novavax.com/news-releases/news-release-details/novavax-confirms-high-levels-efficacy-against-original-and-0

(12) Novavax's Production Problems Are Overblown, Says Analyst ... (n.d.). Retrieved from https://www.nasdaq.com/articles/novavaxs-production-problems-are-overblown-says-analyst-2021-03-29

(13) Pardi, N., Hogan, M. J., Porter, F. W., & Weissman, D. (2018). MRNA vaccines — a new era in vaccinology. Nature Reviews Drug Discovery, 17(4), 261-279. doi:10.1038/nrd.2017.243

(14) Rapaka, R. R., Hammershaimb, E. A., & Neuzil, K. M. (2021). Are some COVID vaccines better than others? Interpreting and comparing estimates of efficacy in trials of COVID-19 vaccines. Clinical Infectious Diseases. doi:10.1093/cid/ciab213

(15) Remmel, A. (2021). COVID vaccines and safety: What the research says. Nature, 590(7847), 538-540. doi:10.1038/d41586-021-00290-x

(16) Suspend AstraZeneca use for people under 55, vaccine committee recommends | CBC News. (2021, March 29). Retrieved from https://www.cbc.ca/news/politics/astrazeneca-under-55-1.5968128

(17) Management of Anaphylaxis at COVID-19 Vaccination Sites. (2021, March 03). Retrieved from https://www.cdc.gov/vaccines/covid-19/clinical-considerations/managing-anaphylaxis.html

Democracy and Legitimacy in the Canadian Senate

Benjamin A. Turner

Introduction

The question of legitimacy in the Senate is as old as the upper chamber itself. The first form of the upper chamber in 1791 was hereditary, this state of affairs triggered the first senate reform in relatively short order, by 1810 Canada had an appointed but no longer hereditary upper chamber (O'Brien, 2019). This was far from the last difficulty the senate experienced. The quest for greater legitimacy in the Canadian Senate appears to be one of our oldest pastimes, but what exactly does a legitimate senate look like? All of the senate reform proposals since 1980 have had a common factor: democracy. Each commission and proposal suggested a different balance for how to reform the senate to be more relevant, more useful, and more legitimate in the eyes of Canadians, but they all identified democracy as an important part of the solution.

In this paper, I will argue that yes, perhaps a democratically elected senate may enjoy more legitimacy with the political body in superficial terms, but this solution affects more than just the appointment process of senators. When considering legitimacy in terms of both the appointment process and its function within government, I find that electing senators grants legitimacy while at the same time potentially undermining its function as a complementary body in the bicameral model of Canadian federalism (Hulme, 2016). Simply put, the answer to the question of whether a democratically elected senate would be more legitimate is: yes, and maybe not.

Defining Legitimacy

There have been many major attempts at senate reform in Canada in the last 40 years, including the Charlottetown and Meech Lake accords, and the 2014 Supreme Court Senate Reference case. None so far have been implemented, but would any of these proposals have fixed the underlying issues of the fact that Canadians lack confidence in their senators? To determine the potential effectiveness of these reform proposals, it is necessary to replace our implicit definition of legitimacy with an explicit one. I have reviewed the work Stillman (1974) to fill in this blank.

Stillman puts it this way: "A government is legitimate if and only if the results of governmental output are compatible with the value pattern of the society." In applying this definition to senate reform, government output is prudently viewed as both the appointment process for senators, as well as how well the senate would likely be to perform its function in government as a chamber of sober second thought. The crucial element of this function within Canadian federalism is that the Senate is not to obstruct or undermine the House of Commons, it is to offer constructive suggestions to improve legislation and then send it back to the elected lower house for final reading.

Reform Attempts

There are eight major reform attempts and committee recommendations that I have reviewed while putting together this paper, including the more recent changes made by Prime Minister Justin Trudeau. The changes made by Mr. Trudeau are unique for a few reasons, notably: he does not call for democracy in the Senate. He dissolved the liberal caucus and established an independent advisory board for the appointment of new senators, granting liberal appointees a level of autonomy not seen in the Senate since 1848-1867 (O'Brien, 2019). He respects the nature of the senate as a complementary body of sober second thought (Hulme, 2016).

And perhaps most importantly, since the Trudeau changes are not subject to the near impossible standard of a constitutional amendment, they have been successfully implemented (Docherty, 2002).

In the autumn of 2013, the government under Prime Minister Stephen Harper submitted a set of reference questions regarding senate reform to the Supreme Court (Hulme, 2016). The legislation Mr. Harper was testing would require every province in Canada to enact senator-in-waiting legislation similar to the one on the books in Alberta at the time. The function of this legislation was to put senate candidates on the ballot any time there was a vacancy for the province in question; citizens of that province would elect a candidate and the Prime Minister would subsequently appoint that individual. Stephen Harper referred this proposed legislation to the Supreme Court because he aimed to use parliament to unilaterally alter the senate, thus side-stepping the issue of passing a constitutional amendment which was the roadblock that sank the Charlottetown Accords and the Meech Lake Accords (Docherty, 2002). The Supreme Court, as previously mentioned, blocked this attempt to bypass constitutional amendment and offered a specific and narrow definition for the function of the senate, primarily focusing very specifically on the complementary function of the Senate and eliminating considerations like regionality and the balance of power.

This judgement effectively killed Prime Ministers Harper's attempt at senate reform. After years of claiming that he could accomplish reform without the need for messy constitutional negotiations, the Supreme Court solidly shut the door on the possibility (Hulme, 2016).

This interpretation of the function of the senate speaks to the matter of legitimacy because it lends absolute clarity to the part the senate is intended to play in Canadian government and firmly establishes the need for a constitutional amendment to alter the role. On the topic of output, if the senate is to be a complementary body to the House of Commons, is it not more likely that elected senators would effectively cease to function as a body of sober second thought and instead become an additional partisan forum with its own perceived mandate to govern? If this is the case it is not difficult to imagine that if the senate finds itself with a different balance of power from the House of Commons, the upper chamber would become an obstructive body or that it might even actively work to undermine the efforts of the elected members of the lower chamber.

For a government to be legitimate, the results of its output must be compatible with the values of the society it governs. At first glance it would appear that an elected senate would be in line with the values of Canadian society. After all, Canada is a democratic nation. I believe this is what makes the concept of an elected senate attractive. It's certainly not a hard sell to the political body, especially when the appointed senate has felt to many as unaccountable through numerous scandals over the years.

But is an elected senate the best way to achieve legitimacy? The function of the senate as a complimentary body to the House of Commons must be taken into consideration; changes that threaten to undermine that role may also threaten the legitimacy of the senate (O'Brien, 2019). The trouble with electing the Senate is that if senators feel they are accountable directly to voters, they feel that they have a mandate of their own. Would creating a second democratic assembly not risk a scenario where one elected body interferes with the next? It creates a scenario where the Senate could see itself as being in competition with the House of Commons instead of being a complimentary body. If the Senate is competing with the lower chamber, it is not difficult to imagine it creating dysfunction within the legislative branch as a whole. That dysfunction would certainly run counter to the values of Canadian society as expressed in the supreme court reference (Hulme, 2016). In considering whether a democratic senate would be in line with the values of Canadian society, we must look at the situation holistically. Balancing the popular will of the people in the moment with the long-established traditions of a society is essential to determining its values. Further research would offer more clarity in this matter.

Conclusion

While the idea of an elected senate, though easy to sell to voters, is not necessarily a better structure. Legitimacy is not automatically the result of democracy; as noted by Stillman, it is important that we do not allow our conclusions to be skewed by ideological affiliations. I fear that this is a trap Canadian leaders have always fallen into with regards to senate reform.

While nearly all of the reform proposals brought forth since 1980 have adopted the idea of an elected senate, it would seem that those that did failed to deeply examine the consequences of that course. It would certainly fix the legitimacy problem evident in the appointment process of senators, but more broadly I am doubtful that an elected senate would be an improvement. Another likely scenario is that it would be a lateral move to a different sort of illegitimacy. We must be wary of trading one devil for another, particularly when, as Prime Minister Trudeau has so successfully demonstrated, there is lower hanging fruit.

References

(1) Bakvis, H. (2000). Prime Minister and Cabinet in Canada: An Autocracy in Need of Reform? Journal of Canadian Studies/ Revue d'études canadiennes, 35:4 (pp. 60-79) University of Toronto Press.

(2) Docherty, D. (2002). The Canadian Senate: Chamber of Sober Reflection or Loony Cousin Best Not Talked About, Journal of Legislative Studies, 8:3 (pp. 27-48) Routledge, Taylor & Francis Group

(3) Hulme, K. (2016). Alberta's Great Experiment in Senatorial Democracy, American Review of Canadian Studies, 46:1 (pp. 33-54) Routledge, Taylor & Francis Group

(4) O'Brien, G. (2019). Discovering the Senate's Fundamental Nature: Moving beyond the Supreme Court's 2014 Opinion, Canadian Journal of Political Science, 52:3 (pp. 539-555) Cambridge University Press

(5) Stillman, P. (1974). The Concept of Legitimacy, Polity, 7:1 (pp. 32-56) University of Chicago Press

False/Inaccurate News Being Spread Online About COVID-19

Benjamin A Turner

While health agencies and governments around the world scramble to contain the spread of a deadly pandemic, the World Health Organization (WHO) warned of an "infodemic" as well. Misinformation about COVID-19 runs wild across the internet, fuelling all manner of ugliness among private citizens who have been taken in by one crackpot theory or another. From allegations that the disease was created in a lab, to vaccine misinformation, or outright denial of the severity of the crisis, there's fake news of all sorts being spread.

This is not a new phenomenon; the internet has always been characterized by how efficiently it spreads fake information. From its earliest days it was a hotbed for conspiracies about the Kennedy assassination and the moon landings. More recently misinformation has been spread on the web not only by isolated scammers or paranoid conspiracy theorists, but by nation states and people in power to meet their goals; in other words, we have seen the weaponization of misinformation for political gain. Everything from fraud claims about the 2020 US presidential election to the Russian sponsorship of vaccine misinformation targeting western nations to sow instability and discontent. These methods have proven shockingly effective and economical.

For COVID-19, there are many variations of the conspiracies. For example, for those arguing that the disease was created in a lab there were initially 2 schools of thought: one camp argued that the Chinese intentionally developed and released the disease to increase their market share in the global manufacturing market, and another camp proposed that China created the disease but that it was released by accident. More recently the theory has become more convoluted, as these things are bound to do, with the theory that the disease was developed in a lab in the United States by an international team of scientists from the US, China, and Germany, among others, and then intentionally released as a global population control measure.

For vaccine denialism we see everything from your standard vaccines cause autism, are dangerous, aren't tested, are just a cash grab for big pharma theories that have been demolishing vaccination rates in the west for 20 years now. But there are others who argue unironically that Bill Gates developed the vaccines as a pretense to inject everybody with a microchip, these people routinely post their arguments from their smartphones which are literal tracking devices and to which most of us are so addicted that we won't even go to the bathroom without them.

Social media platforms like Facebook and Twitter have been the primary vectors for transmission of this infodemic but facing massive public pressure Facebook has taken token steps to combat COVID-19 misinformation. These include banning some popular figures that spread these falsehoods and adding banners with links to public health agencies to any posts mentioning COVID-19. Unfortunately, it is unclear whether Facebook has been too busy counting the ad revenue generated off those false claims to notice the staggering body count those lies have fueled.

Screwballs

Mardon, C. A. (2014). Screwballs. Edmonton: Golden Meteorite Press.
120 pp. ISBN-10: 1897472544

Review by Benjamin A. Turner

Screwballs is a memoire recounting the experiences of Dr. Catherine Mardon, focusing largely on her time practicing law in the United States and the interactions she had with people, some clients and some not, who suffer from various mental illnesses. The book gives a very honest and compassionate look into the struggles of a stigmatized community at the individual level, often presenting the illogical decisions people make followed by an explanation of why those choices seem perfectly reasonable to the person living with their respective mental illness. Dr. Mardon was born to a poor family in Oklahoma, worked her way through university earning a law degree, and made a name for herself representing people who suffered from various forms of oppression, from hostile work environments to family estrangement; she did this right up until suffering her own set of traumatic injuries that effectively ended her law practice.

At the time of its release, mental illness had become a major point of public focus in North America. Dr. Mardon and her husband had already long been advocates for mental health policy, but it seems to have been the ideal time to put these stories into the world. She offers only her personal anecdotes to raise awareness for the struggles of people suffering from mental illness and stands as an example of how to care for the complicated needs of members of this vulnerable community, and also warns of how to avoid being taken advantage of, reminding readers that self care is also important. One hopes that the people who read this book will gain an insight into how to communicate with a diverse spectrum of people, how to withhold judgement, and how to seek counsel from their community when they feel lost.

The accounts offered here are heartbreaking, inspiring, and not without humour. One finds this book to be unbelievable if for no other reason than for the sheer patience and understanding displayed by the author. She is humble in how she recounts these experiences and discusses her struggles to do what she believes is required of her while staying true to her moral compass as informed by the church.

The book begins with a recounting of lessons learned through baseball as a child and links those lessons to the title of the book. The way she tells it, a screwball is a very difficult pitch to master as either a pitcher or hitter in baseball, it is also used to describe eccentrics and anyone who is just not like everybody else, and finally how it is used as a pejorative to describe people with mental illnesses. It goes on to discuss her childhood and family, their struggles and how they were touched by mental illness. She discusses the stresses she saw in her parents' marriage, her fathers post traumatic stress from combat duty in the Second World War, her 6th grade teacher who fell victim to stigma and the associated guilt she felt not being sure if she and her classmates had caused their teachers meltdown.

It is not until later, when she is in college, that she really has a brush with mental illness that will establish a pattern, however. She takes in her nephew when he is struggling and tries giving him a leg up. Being in classes full time and holding down a job to pay her bills, she learns a difficult lesson: you can't save a fish from drowning. This nephew steals, lies, and cheats, generally creating chaos in her life until she is forced to kick him out of her house. This first foray into helping those struggling with addiction and mental illness proves frustrating and heart wrenching, but valuable. It is an important warning to avoid being taken advantage of.

The book goes on to recount her decision to attend law school, and some of her early forays into representing the underprivileged.

She cuts her teeth largely working with patients in the early AIDS epidemic, noting that in the early days there were very few professionals willing to interact with the community. A time governed by fear of the unknown and the chaos that those fears sowed, it seems Dr. Mardon may have been touched by the combination of compassion for the AIDS patients and disdain for the prejudice demonstrated by lawyers and doctors who wanted nothing to do with them.

Working with AIDS patients opens the door to interactions with all sorts of disadvantaged people, most of whom are predisposed to mental illness. These include the homeless, the addicted, those marred by physical injuries that fell through the cracks, and folks struggling with diverse sexual orientation and gender identity issues in a society that is intolerant of the same.

These interactions lead to fighting for insurance claims, welfare benefits, labour disputes, custody battles and more. She takes in a friend with severe dependency issues, a complicated grab bag of mental illnesses, and struggles to keep her on medications and out of jail. The two of them live together for years before the relationship sours, but again Dr. Mardon displays an uncommon patience and generosity of spirit before being driven over the edge, sacrificing a great deal of her own health and wellness in an attempt to care for another.

Readers of Screwballs will gain valuable insights into how to approach mental illness. Perhaps the most valuable piece of information is how to speak to someone with whom you are uncomfortable because they immediately strike one as being abnormal. This book demonstrates how it is often our own response to a situation that can trigger poor outcomes; patience and understanding are essential to avoiding causing new trauma. It also speaks to the value of putting oneself in situations that are uncomfortable, that sometimes it is necessary to take a stand. The best example of this is when she attends the execution of one of her clients for whom she fought to get off death row. Not because he was innocent, but because the death penalty is abhorrent to her. The only things worth doing are the things that are difficult.

It is difficult not to get hooked and read this novel in one sitting, the stories are so fascinating and eclectic. Dr. Mardon's style is easy to read and endearing, she is unpretentious and to the point. One frustration, however with the book is with the overall style. It feels a little unpolished in its structure, particularly with regards to the pacing. It is unclear sometimes when exactly things are happening and at several points the reader finds they have jumped ahead by years on end. She is also succinct to a fault. One may wish for deeper elaboration of many of the anecdotes, things often move so quickly in Screwballs that the reader has hardly any time to absorb the gravity of an event before moving on to the next incredible act of selflessness. Overall Screwballs is an impactful statement on the importance of mental health awareness. It lends to the incredible experiences of an accomplished woman with the patience of a saint and works as an instruction manual for how to help without becoming a martyr, how to be accessible without being exploited. Both a cautionary tale and inspirational piece, readers will be better informed for having read it.

US Rejoins Paris Climate Accord: Too Little, Too Late?

Benjamin A Turner

The election of Joe Biden as President of the United States marks a very different approach to international relations from the US. The new administration has brought back the multilateral model of the United States as a leader on the international stage, quite the reversal from the America first politics of just last year. One major shift in US policy is toward the Paris Climate Accord, which the US formally left late last year under President Trump. The new administration has formally re-entered the pact to limit and reduce harmful greenhouse gasses.

On Twitter US Secretary of State Tony Blinken said, "It's a good day in our fight against the climate crisis, as the United States is once again a Party to the Paris Agreement." He went on, "The work to reduce our emissions has already begun, and we will waste no time in engaging our partners around the world to build our global resilience."

This news will surely be music to the ears of climate activists, but has the recent political instability in the US made them an unreliable ally? Former President Trump was able to undo 70 years of international policy dogma in his one term in office, so it could be understandable for other signatories to the pact to view the US as too unstable to be counted on.

While domestic opponents of the accord are quick to point out that the US is outpaced in emissions by China, it is noteworthy that China has signed the Paris agreement and also that they did not later withdraw, as the US has. And while China is responsible for roughly 29% of global greenhouse gas emissions according to United Nations Framework Convention on Climate Change, making them the largest polluter in the world, the US comes in second place with 14%. While the US has proven fickle on the subject, no serious climate effort can reasonably be achieved without their cooperation. The industrial base of the United States is a huge source of greenhouse gas emissions, and serious changes need to be made there to meet the targets set out in the Paris agreement.

The election of President Biden rings a hopeful note for those who value climate action and international cooperation, but the damage done by the previous administration will be lasting. From policy changes at the EPA to security breaches of intelligence data shared by US allies, a great deal of work lays ahead to restore relationships and regulatory systems. On February 19th, 2021, President Biden said, "This morning, I met with my fellow G7 leaders for the first time as president. I made clear that America is back at the table." While a great deal of damage has been done, the US president appears to be committed to repairing America's damaged relationships abroad; rejoining the Paris agreement is a big step in the right direction. Frustrated though they may be, US allies have no choice but to be glad the US is back at the table.

The Proliferation of Antimicrobial Resistance Due to COVID-19 Sterilization Practices

Paawan Virdi, Alexander Martin, Daivat Bhavsar, Austin Mardon

With the spread of COVID-19 currently being at the forefront of global issues, prevention and treatment procedures are being employed at unprecedented levels. The circumstances created by the spreading infection have yielded conditions in which another health issue—antimicrobial resistance (AMR)—can thrive. The World Health Organization has declared AMR to be amongst the most severe global health threats facing humanity; deeming it a complex problem that necessitates coordinated action (World Health Organization, 2020). The United Nations also warned that in the absence of meaningful action, AMR could escalate to the point of 10 million deaths a year by 2050, and could cause economic crises forcing 24 million people into extreme poverty by the year 2030 (Interagency Coordination Group on Antimicrobial Resistance, 2019). COVID-19 presents a unique international crisis that could exacerbate the current situation by favouring the proliferation of AMR through excessive sterilization and the improper use of antibiotics.

Proliferation of AMR in Clinical Settings

One systemic factor exacerbating the AMR crisis is increased patient hospitalization. The sudden emergence of the COVID-19 pandemic and the lack of preparation has led to greater patient loads. An increased number of patients could lead to a greater number of nosocomial infections, i.e. those occurring due to pathogens that reside in a certain location (Rawson et al., 2020) . An example of this is Methicillin-Resistant Staphylococcus Aureus, or MRSA, which is present in 5% of hospital patients in the United States. Its primary method of spread is patient-to-patient via the hands of healthcare workers (Morell and Balkin, 2010; Centers for Disease Control and Prevention, 2019). Spread of AMR is also possible due to the improper usage of antimicrobials in clinical settings. For instance, the use of antibiotic agents in the treatment of viral diseases. One study that was conducted in intensive care units across 88 countries demonstrated that while suspected or proven bacterial infections were present in 54% of patients, 70% received at least one antibiotic for prevention or treatment (Getahun et al., 2020). This suggests that around 16% of patients received antibiotic treatment in the absence of bacterial infection, thereby contributing to the selective pressure favouring resistance without any benefits to the patient.

Antimicrobials and Misinformation

COVID-19 has also influenced the spread of AMR at the individual level. Being in the world's spotlight, it is inevitable that misinformation can spread as there is a constant flow of new information about the disease. An example of this includes the efficacy of antibiotics against COVID-19. A survey in Australia found that 44% of respondents believed antibiotics were effective in the treatment of COVID-19 (Arshad et al., 2020). As this is not the case, the antibiotics would only serve to increase the selective pressure for AMR in countries where self-medication is possible (Monnet and Harbarth, 2020). Excessive sanitation procedures also increase selective pressure for resistance, and have likely increased in households and long-term care facilities in the past year (Monnet and Harbarth, 2020).

Domestic and Commercial Sterilization Practices and COVID-19
The increased use of disinfectants as a result of the pandemic may also contribute to the proliferation of AMR. The purchasing

of hand sanitizer has surged globally as a consequence of the COVID-19 pandemic, with sales increasing drastically in countries around the globe (Gibson, 2020). The use of sanitizers containing ethanol or some other non-antibiotic substance is often employed in public settings such as commercial buildings, workplaces, and educational facilities. Recent studies have demonstrated that certain pathogens may be developing resistance to such alcohol-based disinfectants. In particular, for the nosocomial infections caused by Enterococcus faecium, samples taken past 2010 demonstrated alcohol tolerance an order of magnitude greater than those of previous isolates (Pidot et al., 2018). Further investigation revealed that the increased resistance is likely associated with mutations in genes involved with carbohydrate metabolism (Pidot et al., 2018). These findings are demonstrative of the potential acquisition of resistance to alcohol based biocidal agents in a manner similar to that for antibiotics (mutagenesis).

Non-alcohol based biocidal disinfectants have also demonstrated the ability to induce antibiotic resistance. For example, triclosan was a common element of antimicrobial soaps in the US until it was banned by the FDA in 2016 over concerns regarding efficacy and resistance (Food and Drug Administration, 2017; Kampf, 2018). However, triclosan is still utilized in such products in other countries, such as Canada. Another example includes benzalkonium chlorides (BACs), which are common elements of antimicrobial soaps, laundry soaps, and disinfectants (Merchel Piovesan Pereira and Tagkopoulos, 2019). The use of BACs in commercial products has been associated with the potential proliferation of microbes that are cross-resistant to multiple biocidal agents. The increased use of disinfectants arising from pandemic practices could potentially increase the selective pressure for microbes with AMR.

Nevertheless, the use of biocidal agents is necessitated as individuals come into contact with each other and new environments. The increased use of disinfectants are mandated as a consequence of the risk of transmission associated with contacting people or surfaces that potentially carry the virus. Against both types of disinfectant, the ultimate sources of AMR are mutagenesis and/or horizontal gene transfer, suggesting that the proliferation of AMR would follow the principles of natural selection (Woodford & Ellington, 2007). Therefore, whether they are alcohol-based or contain broad spectrum antibiotics, their use contributes to selective pressures favouring AMR in microbes. This research further stresses the importance of social distancing practices and limitation of transport. In addition to preventing the use of disinfectants, they also contribute to limiting the spread of resistant infections.

Factors Impeding the Spread of AMR

Nevertheless, it is important to note that unique circumstances have been created, and there may be factors arising from COVID-19 that serve as impediments to the spread of AMR. For instance, social distancing practices serve to both eliminate the risk of transmission and AMR. The increased reliance on social distancing practices as a means of reducing spread would serve to reduce the selective pressure for AMR. In addition, increased hand hygiene without the use of antimicrobials would have similar effects as would decreased travel, and the construction of new facilities that are free of nosocomial infections (Monnet and Harbarth, 2020). The reduced selective pressure for AMR presents methods through which the circumstances created by the COVID-19 pandemic could potentially prevent the spread of AMR.

Conclusion

Multiple factors suggest that the global response to COVID-19 could support the proliferation of AMR. However, society is a complex system. There are also circumstances that have been created which could impede the spread of AMR. For instance, social distancing practices, increased hand hygiene, decreased travel, and the construction of facilities free of nosocomial infections all likely limit AMR spread opportunities (Monnet and Harbarth, 2020). While one can try to predict outcomes with the current scientific body of knowledge, novel circumstances may yield novel outcomes. Ultimately, we are presented with circumstances that can exacerbate or ameliorate the AMR crisis. Time will tell how the preventative measures for the spread of COVID-19 will augment existing health crises such as AMR.

Paawan Virdi, BSc (McMaster University, virdip1@mcmaster.ca) is a third year undergraduate student studying integrated science with a concentration in biology.

Alexander Martin, BA (University of Waterloo, alexander. martinale3@gmail.com) is a fourth year undergraduate student in Honours Rhetoric, Media, and Professional Communication.

Daivat Bhavsar, BSc (McMaster University, bhavsd5@mcmaster. ca), is a third year undergraduate student studying biochemistry.

Austin Mardon (aamardon@yahoo.ca), Ph.D., CM, FRSC is an Assistant Adjunct Professor in the University of Alberta Department of Psychiatry, and the founder of the Antarctic Institute of Canada.

References

Arshad, M., Mahmood, S. F., Khan, M., & Hasan, R. (2020). Covid -19, misinformation, and antimicrobial resistance. *BMJ*, 371, m4501. https://doi.org/10.1136/bmj.m4501

Centers for Disease Control and Prevention. (2019, June 26). *Methicillin-resistant Staphylococcus aureus (MRSA)*. https://www. cdc.gov/mrsa/community/index.html

Food and Drug Administration. (2017, December 20). *Safety and Effectiveness of Health Care Antiseptics; Topical Antimicrobial Drug Products for Over-the-Counter Human Use*. Federal Register. https:// www.federalregister.gov/documents/2017/12/20/2017-27317/ safety-and-effectiveness-of-health-care-antiseptics-topical-antimicrobial-drug-products-for

Getahun, H., Smith, I., Trivedi, K., Paulin, S., & Balkhy, H. H. (2020). Tackling antimicrobial resistance in the COVID-19 pandemic. *Bulletin of the World Health Organization*, 98(7), 442–442A. https://doi.org/10.2471/BLT.20.268573

Gibson, K. (2020, March 2). *Demand for hand sanitizers up 1,400% and sellers are rationing supplies.* https://www.cbsnews.com/news/coronavirus-demand-for-household-cleaners-disinfectants-lysol-clorox-purell-sanitizers-2020-03-02/

Kampf, G. (2018). Biocidal Agents Used for Disinfection Can Enhance Antibiotic Resistance in Gram-Negative Species. *Antibiotics*, 7(4), 110. https://doi.org/10.3390/antibiotics7040110

Pereira, B. M. P., & Tagkopoulos, I. (2019). Benzalkonium Chlorides: Uses, Regulatory Status, and Microbial Resistance. *Applied and Environmental Microbiology*, 85(13). https://doi.org/10.1128/AEM.00377-19

Monnet, D. L., & Harbarth, S. (2020). Will coronavirus disease (COVID-19) have an impact on antimicrobial resistance? *Eurosurveillance*, 25(45). https://doi.org/10.2807/1560-7917.ES.2020.25.45.2001886

Morell, E. A., & Balkin, D. M. (2010). Methicillin-Resistant Staphylococcus Aureus: A Pervasive Pathogen Highlights the Need for New Antimicrobial Development. *The Yale Journal of Biology and Medicine*, 83(4), 223–233.

Pidot, S. J., Gao, W., Buultjens, A. H., Monk, I. R., Guerillot, R., Carter, G. P., Lee, J. Y. H., Lam, M. M. C., Grayson, M. L., Ballard, S. A., Mahony, A. A., Grabsch, E. A., Kotsanas, D., Korman, T. M., Coombs, G. W., Robinson, J. O., Silva, A. G. da, Seemann, T., Howden, B. P., … Stinear, T. P. (2018). Increasing tolerance of hospital Enterococcus faecium to handwash alcohols. *Science Translational Medicine,* 10(452). https://doi.org/10.1126/scitranslmed.aar6115

Rawson, T. M., Moore, L. S. P., Zhu, N., Ranganathan, N.,

Skolimowska, K., Gilchrist, M., Satta, G., Cooke, G., & Holmes, A. (2020). Bacterial and Fungal Coinfection in Individuals With Coronavirus: A Rapid Review To Support COVID-19 Antimicrobial Prescribing. *Clinical Infectious Diseases*, 71(9), 2459–2468. https://doi.org/10.1093/cid/ciaa530

Woodford, N., & Ellington, M. J. (2007). The emergence of antibiotic resistance by mutation. *Clinical Microbiology and Infection*, 13(1), 5–18. https://doi.org/10.1111/j.1469-0691.2006.01492.x

Saving Nature the Unnatural Way: Synthetic Biology and its Applications to Climate Change

Paawan Virdi & Austin Mardon

It should be of no surprise that the industrial-scale farming of animals is a major source of greenhouse emissions. Animal agriculture manages to challenge the combined emissions of effectively all major modes of transportation. Because of this, the United Nations Environment Programme has declared the greenhouse outputs of animal agriculture a problem that must be addressed in order to fulfil the Paris Agreement—a series of goals to decrease the effects of climate change (United Nations Environment Programme, 2018). In addition to releasing an enormous amount of emissions, animal agriculture also uses immense amounts of freshwater, with 92% of humanity's freshwater footprint coming from agriculture, and meat contributing disproportionately (Gerbens-Leenes, Mekonnen and Hoekstra, 2013). While the consumption of meat and other animal derived products is a widespread practice all over the world, it is clear that it comes with hefty consequences. As is often the case with slow-moving environmental crises, the primary solution is one that is quite simple but incredibly difficult to employ: changes in lifestyle. However, through the use of synthetic biology to create lab-grown meats, the transition away from heavy animal agriculture may not be as hard as it seems.

Humanity is often reluctant to sacrifice quality of life for the collective good. The question must be asked: how well do our intentions fair against our willingness to act? Ethics in the realm of science can be hazy. For a field sounding Frankenstein-esque, synthetic biology has potential applications to better the lives of humans across the globe. This goes past enriching the food we eat, as exemplified by the controversial golden rice. Synthetic biology has implications regarding the process of agriculture itself. It works to concurrently address crises of climate change and world hunger. In an article in Forbes magazine, it was cited that the production of synthetic meat could drastically reduce the amount of greenhouse gases produced by agriculture by upwards of 96% (Marr, 2019). As the search for renewable energy sources continues.

Meat-substitutes are nothing new, Beyond Meat has been around for a while and is an option you have probably encountered before. However, advancements in synthetic biology have the potential to take it to the next level. In an article by the Guardian journalist George Monbiot, he details his experiences consuming flour made from multiplying bacteria found from the soil. The resulting flour was then used to create an artificial pancake which tasted "just like a pancake" (Monbiot, 2020). While the methods of accomplishing this are varied, one involves culturing stem cells from animal muscle tissue to effectively grow meat in a petri dish—it is meat in vitro (Ireland, 2013). Although such technology has much room for development, the prospect appears to be promising. Another notable example is the production of a protein dubbed "solein" by the company Solar Foods. The process mimics that used by plants, creating food from external energy and nutrients. The process involves fermenting microbes in a growth medium in which compounds such as hydrogen, carbon dioxide, and nutrients that plants normally absorb through soil are supplied. Energy to power this process comes from electricity gathered through renewable sources. This process is analogous to photosynthesis, producing protein rich cells similar to dried soy instead of carbohydrates. The product is then intended to be used as a protein component of foods, and could potentially be expanded to contribute to the growth of real meat cells (Solar Foods, 2019). As amazing as this procedure already is, it also comes with environmental benefits.

When compared to the production of beef, the production of solein utilizes 700 times less water. Along with utilizing less water, when it comes to the amount of food produced, solein also utilizes less land area, produces fewer greenhouse gases, and is carbon negative (Solar Foods, n.d.).

So we have the potential to make all of our food in ways that could drastically reduce the emissions of one of the biggest contributors to climate change, so what? All the rhetoric of the alternatives changing the same may not be sufficient in order to actually get people to eat the foods. If the taste does not act as a barrier to people consuming it, then perhaps the method of production. Despite the fact that genetic modification can have nutritional and economic benefits, it has still bred controversy. If everyone consumed its products, could synthetic biology prove an effective avenue for addressing the climate crisis? Perhaps, but given ideal conditions, so could many other approaches to climate change. The common theme is abiding by the necessary lifestyle changes. Success with synthetic biology is not certain, but humanity must learn from precedents set by carbon taxes and other attempts at curbing emissions. We can try to make it easier on ourselves, but at the end of the day, changes to the way we live in one way or another are necessary.

Paawan Virdi, BSc (McMaster University, virdip1@mcmaster.ca) is a third year undergraduate student studying integrated science with a concentration in biology. Austin Mardon (aamardon@ yahoo.ca), Ph.D., CM, FRSC is an Assistant Adjunct Professor in the University of Alberta Department of Psychiatry, and the founder of the Antarctic Institute of Canada.

References

Gerbens-Leenes, P.W., Mekonnen, M.M. and Hoekstra, A.Y., 2013. The water footprint of poultry, pork and beef: A comparative study in different countries and production systems. *Elsevier B.V.*, 1–2, pp.25–36.

Ireland, T., 2019. *The artificial meat factory - the science of your synthetic supper.* [online] BBC Science Focus Magazine. Available at: <https://www.sciencefocus.com/future-technology/the-artificial-meat-factory-the-science-of-your-synthetic-supper/> [Accessed 20 Apr. 2021].

Marr, B., 2019. *The Future of Food: Amazing Lab Grown And 3D Printed Meat And Fish.* [online] Forbes. Available at: <https://www.forbes.com/sites/bernardmarr/2019/06/28/the-future-of-food-amazing-lab-grown-and-3d-printed-meat-and-fish/#1ce33b2b46f6> [Accessed 17 Apr. 2021].

Monbiot, Geroge, 2020. *Lab-grown food is about to destroy farming – and save the planet.* [online] Available at: <https://www.theguardian.com/commentisfree/2020/jan/08/lab-grown-food-destroy-farming-save-planet> [Accessed 21 Apr. 2021].

Solar Foods, 2019. *Solein® Q&A.* Available at: <https://solarfoods.fi/wp-content/uploads/2019/11/Solein-Q_and-A_FULL.pdf>.

Solar Foods. (n.d.). *Food out of thin air.* Retrieved April 29, 2021, from https://solarfoods.fi/

United Nations Environment Programme, 2018. *Tackling the world's most urgent problem: meat.* [online] United Nations Environment Programme. Available at: <https://www.unenvironment.org/news-and-stories/story/tackling-worlds-most-urgent-problem-meat> [Accessed 16 Apr. 2021].

COVID-19 and its Differential Impacts on Various Social Groups in North America

Paawan Virdi & Austin Mardon

Since the beginning of the pandemic in early 2020, it has been evident that different social groups are disproportionately affected by the COVID-19 pandemic in accordance with their race and class. Racial and ethnic minorities have been shown to be especially susceptible to the disease. In addition, one's economic status has also shown to be a factor in how prone they are to COVID-19. Various studies have investigated the relationship between one's social categorizations and their vulnerability to COVID-19.

According to the United Nations, indigenous peoples worldwide have shown to be more prone to infection due to the relative inaccessibility of healthcare, prevalence of disease, and systemic factors. Examples of the latter include the lack of access to effective sanitation, the understaffing of local medical facilities, among many others. To top it off, indigenous peoples are also presented with poorer medical outcomes due to discrimination in clinical environments. There are also factors specific to the organization of indigenous societies that make them more prone (United Nations, 2020). For instance, the occurrence of gatherings and multigenerational housing that includes endangered elders. The 2009 H1n1 influenza had a four times greater mortality rate for Native Americans when compared to the general population of North America (Burki, 2021). Similar trends were observed with COVID-19. Native Americans were shown to have much higher rates of disease and mortality (McLernon, 2021). The Government of Canada has taken preventative measures by offering vaccines earlier to some indigenous peoples. As of late April 2021, over 660 indigenous communities across Canada have been in the process of undergoing vaccination (Government of Canada, 2021).

Other racial groups also appear to be relatively vulnerable to COVID-19. In the United States, non-hispanic black people, hispanic people, and latino people have shown hospitalization rates almost 5 times greater than that of white people. While there is no data that suggests such individuals are more genetically prone to the disease, they are more likely to have other conditions which can exacerbate the effects of COVID-19 (Marshall, 2020). This is also true with Native American peoples. In addition, there are a wide variety of systemic factors that have shown to result in the differential vulnerability to COVID-19. For instance, people of color are more likely to be essential workers. According to the Centers for Disease Control and Prevention, in the United States, the proportion of hispanic black or African American workers that are service workers is over 56% greater than the corresponding proportion of white workers (Marshall, 2020). Factors such as geographical location, family and home size, transportation methods, all result in people of color being more vulnerable to COVID-19.

Ultimately, much of the systemic factors that are afflicting some racialized communities are associated with gaps in the distribution of wealth. When the government denies its essential workers the paid sick days they need, those in a less economically stable position cannot afford to stay home. This drastically increases their likelihood of getting infected, simply because they are not in comparable financial positions. In addition, individuals with less wealth are less likely to have higher quality education due to the many financial barriers associated with pursuing further studies and the need to work alongside this. As a result, they are less likely to be in careers where working from home is an option, increasing the risk of encountering COVID-19 in the workplace. Of course even with comparable education, discrimination in other aspects of society function as obstacles to financial mobility (CDC, 2020). COVID-19 has also shown to be much more detrimental to individuals living in poverty, as mortality rates are seen to be much higher amongst poorer groups (Whitehead et al., 2021). The reasons behind this are similar to those mentioned earlier for certain racialized communities: the inability to stay home, to access necessary resources, and to rest and isolate.

COVID-19 has also facilitated the strengthening of prejudice against certain groups. One example of which is the city of Brampton, Ontario. The city consists mostly of visible minorities, and systemic factors such as the large essential worker population and insufficient health support have made the city ideal for facilitating the spread of COVID-19. Nevertheless, many individuals have placed much of the blame for the spread of COVID on the city's large South Asian population, who they believe are disproportionately disregarding lockdown procedures.

Overall, it is evident that the pandemic has struck certain social groups with much greater force than others. Some racialized communities and those living in poverty are at increased risks. In addition, it is important to note how these groups often overlap and that there are common factors that result in the differential outcomes discussed. Understanding that there are systemic factors that render certain groups more vulnerable enables action to be taken accordingly, preventing COVID-19 from exacerbating the circumstances that facilitate inequality in our society.

Paawan Virdi, BSc (McMaster University, virdip1@mcmaster.ca) is a third year undergraduate student studying integrated science with a concentration in biology. Austin Mardon (aamardon@yahoo.ca), Ph.D., CM, FRSC is an Assistant Adjunct Professor in the University of Alberta Department of Psychiatry, and the founder of the Antarctic Institute of Canada.

References

Burki, T. (2021). Covid-19 among American Indians and Alaska natives. The Lancet Infectious Diseases, 21(3), 325–326. https://doi.org/10.1016/S1473-3099(21)00083-9

CDC. (2020, February 11). Community, work, and school. Centers for Disease Control and Prevention. https://www.cdc.gov/coronavirus/2019-ncov/community/health-equity/race-ethnicity.html

Government of Canada. (2021, April 9). Indigenous communities with COVID-19 vaccination underway [Notice]. https://www.sac-isc.gc.ca/eng/1617991515859/1617991548784

Kerr, J. (2020, November 13). Why Brampton has become a hot spot for COVID-19. https://www.theglobeandmail.com/canada/article-why-brampton-has-become-a-hot-spot-for-co vid-19/

Marshall, W.F. (2020, Aug 13)Why is COVID-19 more severely affecting people of color? Mayo Clinic. Retrieved April 30, 2021, from https://www.mayoclinic.org/diseases-conditions/coronavirus/expert-answers/coronavirus-infecti on-by-race/faq-20488802

McLernon, L. M. (2021, April 9). Reports detail high COVID-19 burden in Native Americans. CIDRAP; Center for Infectious Diseases Research and Policy. https://www.cidrap.umn.edu/news-perspective/2021/04/reports-detail-high-covid-19-burden-nat ive-americans

United Nations. (2020, March 30). Covid-19 and indigenous peoples | United Nations for indigenous peoples. https://www.un.org/development/desa/indigenouspeoples/covid-19.html/

Whitehead, M., Taylor-Robinson, D., & Barr, B. (2021). Poverty, health, and covid-19. BMJ, 372, n376. https://doi.org/10.1136/bmj.n376

Historical Precedent, Ideological Movements, and a Culture of Hate

Minahil Syed, Peter Anto Johnson, John Christy Johnson

The residual influence of Puritanism and Whiggery never truly simmered down; while the foundation of America is egalitarian-individualist, the superstructure and ethos are undoubtedly classically liberal. It is tapped into by conservative movements whenever fundamental values are debated, albeit, defended under the guise of 'traditionalism' or 'utilitarianism'. Nonetheless, this is not to feign that classical liberalism is interchangeable with the xenophobia and racism spouted under Trump's presidency.

Politics is not necessarily divisive but values that permeate therein are. To elucidate, classical liberalism merely supports free-market capitalism, division of church and state, non-interventionism, traditional social norms, and the consecration of rights in accordance with the Declaration of Independence and the U.S Constitution. In essence, classical liberalism is value-neutral and simply a political ideology with an emphasis on economic freedom and the appropriation of positive liberties ("Liberalism," 1996). Hence, resounding statements such as, "I hate x group of people," are not a political stance, they represent a value, a racist value; such which have shrouded themselves under the guise of being a political statement. People have mistaken hatred for ideology.

Thus, dialogue that refers to political values and conservative values as interchangeable **erroneously** reduces rightist beliefs to the political level while sanctifying leftist beliefs as morally neutral. This paradigm shift has not always been as transparent as it currently is, it is a period borne as a by-product of how the government's head (prime minister, president, etc.) and their respective identity shape the country's political image. The breakdown of anachronistic systems of international relations gave way to an interrelatedness resultant of globalization, wherein the state is no longer the sole institution imposing on its citizenry, it has become one of many actors within the domestic and international sphere (Hurrell, 2020).

In 1991, David Duke, ex-Grand Wizard of the Klu Klux Klan, neo-Nazi, and blatant white supremacist, ran for governor in Louisiana. The nonpartisan blanket primary allowed all candidates, irrespective of their party, to run for candidacy. Duke served as a representative on behalf of the Republican party and amassed 31.71% of votes (491,342) during the 1st gubernatorial election. He placed second only to Edwin Edwards, a Democrat candidate who accumulated 33.76% or 523,096 of votes. Duke's placement invoked national attention and condemnation as his past resurfaced, however, when primaries began, not many provided sufficient attention to his background.

Duke was a talented orator, and his campaigning was contingent on pressing issues for a downtrodden state in need of dire aid. Hence, his positioning and outcry on such concerns allowed him to flawlessly shield and divert attention from his prolific anti-Semitism and history of racial code (Maraniss, 1991). He campaigned against higher taxes, for a weaker welfare state, and opposed affirmative action. Louisiana's demographics were overbearingly white, with a working-class who had suffered in the recent recession and the Oil Bust of the 1980s; the higher unemployment rates relative to the national average only further exacerbated tensions (Bridges, n.d.). Duke's agenda was ridden with racial undertones, he attacked affirmative action and the welfare state as mechanisms that catered to the poorest strata, and his platform was well-received by the majority white district irrespective of their social standing and income level (Serwer, 2017).

In 1989, Duke appeared at an anti-tax rally in Metairie, Jefferson Parish, opposing a tax reform package which would relieve the burden from businesses, consequently increasing income taxes while simultaneously decreasing sales tax ("Louisiana Tax Bills Passed, But Revenue Woes Remain," 1989). Although the package sufficed as an attempt to make the state appear more inviting to businesses, tourists, and investment, it inevitably shifted the onus of responsibility on middle-class voters to fund (Bridges, n.d.).

With Duke leading the opposition, he depicted himself as the invisible hand for voters. His platform resembled that of his present-day caricature, Donald Trump. Duke's popularity could have been attributed to the mixture of his personal and political vendettas. He rallied against issues that appeal to a predominantly homogenous electorate, using marginalized communities as a scapegoat for his attacks, and he capitalized on the result therein to advance his white supremacist agenda (Maraniss, 1991). Sound familiar?

From the onset of Obama's presidency, conspiracy theories circulated regarding the birthplace of the Hawaiian-born candidate colloquially dubbed the "birther movement," which made frequent impasses towards his race, religious beliefs, and citizenship status to assert that he was ineligible to serve as President of the United States.

Birther rhetoric relies on the logic of 'othering' and perpetuating foreignness to reinforce white nationalist sentiment (Pham, 2015). Conservative news outlets such as the National Review Online circulated controversial think-pieces concerning Birther rumours amongst mainstream channels, only further intensifying the controversy (Pham, 2015). After the eventual release of Obama's birth certificate, both the Republican and Tea Party distanced themselves from the Birther movement, yet capitalization upon the inherently xenophobic and anti-Black discourse did not end there (Zelizer, 2021). Donald Trump remained a prominent voice within the movement for years, insisting too that President Obama was a Muslim rather than Christian (Abramson, 2016).

The fabrication that the President was not born 'domestically' attacked the authority of America's first Black President, whereas the insistence that he is Muslim debased his 'priorities' and sufficed as a mechanism to garner votes - intrinsic on fear of Muslims and Islam.

American mobilization and rhetoric throughout the eras have been contingent on relative gains, forwarding their own zero-sum agenda, and the ideological stance of the ruling political figurehead. During Trump's presidency, he attacked the proliferation of 'fake news' (i.e. information which discredited or overturned his dogma), condemned China on an international stage (resulting in a relative increase in xenophobic hate crimes and messages domestically), referred to Mexican and Muslim immigrants using prejudiced stereotypes (in racially-coded speeches to appeal to his constituencies), called Black Lives Matter protestors 'terrorists,' and blatantly refused to condemn white supremacy (Dewan, 2020). Trump's statements were never value-neutral, they are similarly riddled with racist undertones, and much like Duke, Trump took advantage of a fractured electorate and convincingly made it his campaign motto.

Human rights are not politics, *not all politicized things constitute politics.*

The Trump administration has abandoned core ideals fundamental to the policymaking sphere of the United States, such being the creation of multilateral agreements, alliances, abidance with international law, engaging in multipolarity, environmental protection (via climate change denial), and the protection of human rights. Instead, virulent far-right propaganda has shaped both domestic and foreign policy.

An ascribed populist, Trump has conjured up a 'global elite' who jointly manipulate against the working class, which allowed him to fuel anti-immigration rhetoric and subsequently impose stricter immigration controls, thereby painting the American people as victims of a global anti-worker agenda (Lacatus, 2020). Trump has led to a corrosion of liberal internationalism contingent on the victimization of the 'true' American people, who believe that their detriment is resultant of immigration and globalization.

Trump has further showcased his subliminal agenda via his use of naming and language; in dubbing the Coronavirus, the 'Chinese virus' or 'foreign,' he displayed a deliberate attempt at othering and displacing responsibility. When questioned during a press conference regarding his usage of the term, Trump responded, "It's not racist at all. No. Not at all. It comes from China. That's why. It comes from China. I want to be accurate" ("Press Conference: Donald Trump Joins the Daily Coronavirus Pandemic Briefing - March 18, 2020," 2020). While Trump's intentions may not have been racist, his language and prose have unquestionably contributed to the proliferation of anti-Asian hate crimes domestically. The Department of Homeland Security themselves issued a statement concerning the exploitation of the pandemic by 'violent extremists' and nationalists seeking to exploit growing unease to further their ideologies and intimidate victims (Mallin & Margolin, 2020).

The personification of the virus associated the pandemic with ethnicity; creating an alarmist narrative that prompted questions about whether 'foreigners' could be trusted; which only further fed into Trump's anti-immigrant rhetoric and racism (Viala-Gaudefroy & Lindaman, 2020). The reason why Trump initially denied the virulence of COVID-19 was to avoid hindering chances at re-election when it seemed impossible to no longer deny, he turned Asian-Americans into scapegoats.

In May 2020, following the murder of George Floyd, Black Lives Matter protests began anew to combat the racially charged murders of Black men and women resultant of police brutality. Nationwide marches resulted in arrests, tear-gassing of innocent protestors, and even the displacement of the National Guard (Chavez, 2021). Trump referred to protestors as 'thugs, anarchists, and terrorists,' however, oddly enough, this description was not extended to the neo-Nazis and alt-right white-supremacists who stormed the Capitol building following Trump's incumbent loss against Biden in the 2021 Presidential Election. The belief that the election was 'rigged' or 'stolen' was reinforced by Trump's variable claims of voter fraud and a 'fraudulent election' (Chaggaris, 2020). Trump supporters scaled the building, broke through metal fences, assaulted Capitol police, eventually leading to the evacuation of congressional leaders, reporters, and lawmakers within the building (Chavez, 2021). After hours of carnage, Donald Trump posted a video on Twitter, the transcript reading,

"...This was a fraudulent election, but we can't play into the hands of these people. We have to have peace. So go home. We love you. You're very special. You've seen what happens. You see the way others are treated that are so bad and so evil. I know how you feel, but go home, and go home in peace" ("TRANSCRIPT | 'Go Home': Trump Tells Supporters Who Mobbed Capitol To Leave, Again Falsely Claiming Election Victory," 2021).

Not once did Trump condemn their violence and this is what the terrorists depended on. The extremists attacked **knowing** that the insurgency would play out in their favour following the disingenuous arrest of white terrorist 'Kyle Rittenhouse'[1] (Ray, 2021). Their boldness was a result of precedent that has reinforced time and time again, validating for them that their racism was acceptable, that it is a political stance.

Although recent political happenings are more blatant examples showcasing how entrenched systemic racism and white nationalist sentiment are within American society, the experiences for White Americans with law enforcement are contingent on a lifetime of privilege (Ray, 2021). Trump's directives and foreign policy emphasize how a state's behaviour is contingent on the representative figurehead of the nation, the fluidity of a national culture, and how hegemony is restructured upon confusion between patriotism and nationalism. The result is a teleological increase in racially prejudiced and nationalistic rhetoric domestically, fuelling ignorance and xenophobia which previously remained hidden. The present has become a chapter in American history marked by how cultural individualism has given rise to a cultural ignorance despite the numerous permeating ideologies.

[1] *Kyle Rittenhouse, 17, drove from Illinois to Kenosha in order to murder Black Lives Matter protestors in Wisconsin, he confidently walked past and alongside officers while bolstering an AR-15 and announcing aloud that he had shot people (Litke, 2020). Rittenhouse killed two and wounded a third.*

Bibliography

Abramson, A. (2016, September 16). How Donald Trump Perpetuated the 'Birther' Movement for Years. Retrieved April 26, 2021, from https://abcnews.go.com/Politics/donald-trump-perpetuated-birther-movement-years/story?id=42138176

Boghani, P. (2020, January 13). Racism in the Era of Trump: An Oral History. Retrieved April 26, 2021, from https://www.pbs.org/wgbh/frontline/article/racism-in-the-era-of-trump-an-oral-history/

Bridges, T. (n.d.). The Duke Dilemma. Retrieved April 26, 2021, from https://64parishes.org/the-duke-dilemma

Budowsky, B. (2016, September 16). Colin Powell Is Right: Birtherism Is Racism. Retrieved April 26, 2021, from https://observer.com/2016/09/colin-powell-is-right-birtherism-is-racism/

Chaggaris, S. (2020, November 25). Donald Trump still believes US election was 'rigged'. Retrieved April 26, 2021, from https://www.aljazeera.com/news/2020/11/25/donald-trump-still-believes-us-election-was

Chavez, N. (2021, January 10). Rioters breached US Capitol security on Wednesday. This was the police response when it was Black protesters on DC streets last year. Retrieved April 26, 2021, from https://www.cnn.com/2021/01/07/us/police-response-black-lives-matter-protest-us-capitol/index.html

Dewan, A. (2020, July 26). Trump is calling protesters who disagree with him terrorists. That puts him in the company of the world's autocrats. Retrieved April 26, 2021, from https://www.cnn.com/2020/07/25/politics/us-protests-trump-terrorists-intl/index.html

Hurrell, A. (2020). Rising powers and the emerging global order. In 1336610523 980090656 J. Baylis (Ed.), The Globalization of World Politics (8th ed., pp. 1-617). Oxford University Press.

Retrieved April 26, 2021, from https://bookshelf.vitalsource.com/books/9780192559586

Liberalism. (1996, November 28). Retrieved April 26, 2021, from https://plato.stanford.edu/entries/liberalism/#ClaLib

Litke, E. (2020, August 29). Fact check: Police gave Kyle Rittenhouse water and thanked him before shooting. Retrieved April 26, 2021, from https://www.brookings.edu/blog/how-we-rise/2021/01/12/what-the-capitol-insurgency-reveals-about-white-supremacy-and-law-enforcement/

Louisiana Tax Bills Passed, But Revenue Woes Remain. (1989, March 15). Retrieved April 26, 2021, from https://www.edweek.org/education/louisiana-tax-bills-passed-but-revenue-woes-remain/1989/03

Mallin, A., & Margolin, J. (2020, March 24). Homeland Security warns terrorists may exploit COVID-19 pandemic. Retrieved April 26, 2021, from https://abcnews.go.com/Politics/homeland-security-warns-terrorists-exploit-covid-19-pandemic/story?id=69770582

Maraniss, D. (1991, November 10). DUKE'S OBSESSION: WHITE SUPREMACY WITH A PLAN. Retrieved April 26, 2021, from https://www.washingtonpost.com/archive/politics/1991/11/10/dukes-obsession-white-supremacy-with-a-plan/77100ba3-3c59-434c-9f49-d3db9a573834/

Pham, V. N. (2015). Our Foreign President Barack Obama: The Racial Logics of Birther Discourses. Journal of International and Intercultural Communication, 8(2), 86-107. doi:10.1080/17513057.2015.1025327

Press Conference: Donald Trump Joins the Daily Coronavirus Pandemic Briefing - March 18, 2020. (2020). Retrieved April 26, 2021, from https://factba.se/transcript/donald-trump-press-conference-coronavirus-briefing-march-18-2020

Ray, R. (2021, January 12). What the Capitol insurgency reveals about white supremacy and law enforcement. Retrieved April 26, 2021, from https://www.brookings.edu/blog/how-we-rise/2021/01/12/what-the-capitol-insurgency-reveals-about-white-supremacy-and-law-enforcement/

Serwer, A. (2017, November 20). The Nationalist's Delusion. Retrieved April 26, 2021, from https://www.theatlantic.com/politics/archive/2017/11/the-nationalists-delusion/546356/

TRANSCRIPT | 'Go Home': Trump Tells Supporters Who Mobbed Capitol To Leave, Again Falsely Claiming Election Victory. (2021, January 6). Retrieved April 26, 2021, from https://www.wbur.org/news/2021/01/06/go-home-trump-supporters-us-capitol-transcript

Viala-Gaudefroy, J., & Lindaman, D. (2021, April 13). Donald Trump's 'Chinese virus': The politics of naming. Retrieved April 26, 2021, from https://theconversation.com/donald-trumps-chinese-virus-the-politics-of-naming-136796

White, L., Jr. (2017, April 16). David Duke: Grandfather of the Alt-Right. Retrieved April 26, 2021, from https://www.bayoubrief.com/2017/08/16/david-duke-grandfather-of-the-alt-right/

Zelizer, J. E. (2021, January 12). What Impeachment Won't Change: How the GOP Became the Party of Trump Over Several Decades. Retrieved April 26, 2021, from https://time.com/5928937/trumpism-deeply-rooted-republican-party/

The Breakdown Between Legislation and Rights

Minahil Syed, Peter Anto Johnson, John Christy Johnson

Government institutions, religious orientation, historical precedence, and foundational ideologies are the constituent elements that differentiate the culture of the United States from Canada (Nesbitt-Larking & Adams, 1991, p. 28).

During the American Revolution in the late 18th century, Canada was constituted by a collection of British colonies, which rapidly expanded via the inheritance of United American Loyalists who deferred to the monarchy (Nesbitt-Larking & Adams, 1991, p. 23). The Loyalists brought considerable touches of British conservatism and dogmas stipulating the significance of collectivism to Upper Canada (Wiseman, 2007, p. 23). Amidst the disorder of establishing ideology, American revolutionaries, "Whigs," who later identified as Patriots, and their forebearers, "The Puritans," left Britain in search of freedom to exercise their practice and religion (Baracskay, 2009). The colonies were dominated by Puritans, nonetheless, were not entirely monolithic, distinct sects existed alongside one another. Religious friction during this period drove the Whigs to separate church from state; an approach which has warranted higher rates of belief relative to Canada (Wiseman, 2007, p. 26).

The ideological development of Canada in a conservative direction is resultant of the dogma which arrived at the behest of settlers who carried this philosophy with them to the New World. The emancipation of Loyalists solely rationalizes the transformation of American ideology in a liberal direction and the respective permeation of conservative principles in English Canada (Horowitz, 1966, p. 156).

Canada in terms of Case Law and Discrimination

Historical perspectives have predominantly converged on the plights of English Protestants and French Catholics, with a tendency to ignore the religious and cultural qualms of other minorities.

In Quebec, former Premier Maurice Duplessis repeatedly used his platform to openly discriminate against Jehovah's Witnesses (JW), for instance, arresting them for the distribution of religious pamphlets and literature (Bowal, 2012). Eventually going as far as to once revoke a liquor license (via an ordinance to the Liquor Commission to act on his authority) from Frank Roncarelli, an active JW, critic of the Catholic Church, and of Quebec's general antagonism and marginalization of the JW minority. Roncarelli was often found to be bailing out JW who were arrested (resultant of a Montreal bylaw which required a license for peddling) (Bowal, 2012). In spite, Duplessis revoked Roncarelli's license due to his outspoken support for Witnesses and payment for their bail bonds.

Roncarelli v. Duplessis served as a landmark decision within the Supreme Court of Canada, wherein the Premier was ordered to pay damages to Roncarelli resultant of the religious discrimination to which he had been subjected ("Roncarelli v. Duplessis"). The case established precedent for the defense of religious freedom **and** practice in Canada and as a leap beyond the 'two-nations' paradigm, capturing the diversity within religious communities in the nation as a whole. Following persecution of Jehovah's Witnesses in Quebec during the 1950s, intense lobbying within Parliament allowed Prime Minister John Diefenbaker to pass the Bill of Rights in 1960. The marginalization of JW was likewise cited by PM Pierre Elliot Trudeau as his reasoning for intense lobbying concerning repatriation of the Constitution in 1982;

which was eventually accomplished by adjoining the Charter ("An Institutional History of Religious Freedom in Canada").

Canada as a Secular State and Historical Precedent

In the Pre-Confederation period (1759-1867), Great Britain attempted to institute the Anglican Church in Canada and simultaneously apply anti-Catholic measures, however, noting that their Catholic subjects were forsaking the British empire for acceptance from the American colonists, overturned this measure to dissuade disunion (Jukier & Woehrling, 2010, p. 159).

The basis for these policies was underwritten by the Royal Proclamation of 1763[2] which established the philosophy by which the North American territories[3] would be presided over (Hall, 2006). The Proclamation failed because the French refused to abandon their culture, tradition, and language. Canadian demographics during this period showcased that the majority of immigrants were French rather than British-English, and because the skew of power was in favour of the former, the Proclamation was retracted in favour of the Quebec Act[4] of 1774 (Glover, 2020).

In 1851, the Freedom of Worship Act was adopted, which forbade restrictions on the "...exercise and enjoyment of religious profession and worship, [individuals should be free to practice] without discrimination or preference..." ("Freedom of Worship Act", 2020). Following enactment of the legislation, in 1854, a statute was established to abolish the fiscal and material benefits that the Anglican Church had unilaterally accrued and been granted (Jukier & Woehrling, 2010, p. 159). On July 1st 1867, The British North America Act (BNA) unified Canada under one dominion and

---

[2] *The Royal Proclamation would force the French to abandon their law for the British style; it was an attempted assimilation. The Proclamation made it difficult for French peoples to obtain government jobs. Most Francophones identified as Catholic, whereas most English self-identified as Protestant. In order to obtain a government job, French-Catholics would have to assert loyalty to the monarchy, which they refused to do (Glover, 2020).*

[3] *The territories were abandoned by the French (via the Treaty of Paris which ended the Seven Years War); following the denouement of the war, the North American territory was entrusted to the British (Hall, 2006).*

[4] *Passed in order to gain allyship from the French who were living in Quebec.*

resulted in the creation of the Constitution (which then outlined the distribution of power and jurisdiction between the Parliament and provincial legislatures) (McConnell, 2006). Canada's Bill of Rights was enacted in 1960, however, since it was not part of the Constitution, could easily be changed; following patriation[5] from the British Crown, the Charter of Rights and Freedoms was added to the Canadian Constitution in its entirety (Paradis & Karbani, 2017). The Constitution did not initially contain any guarantee or protection for religious freedoms; the fundamental freedoms within the Charter were what secured these later on (added in 1982).

In the 1867 Constitution, there was no state religion and ecclesiastical matters received no interference from the state, as is the case contemporaneously. Nonetheless, Section 93 of the BNA act asserted provisions for separate denominational schools for both Protestant and Catholic minorities so as to cater to religious difference and pluralism ("An Institutional History of Religious Freedom in Canada", 2020). Furthermore, no financial support was provided for churches, nor did the state collect taxes for the sake of redistribution to support religious communities; institutions of worship were maintained privately, once again highlighting the separation of civil and ecclesiastical affairs in Canada (Jukier & Woehrling, 2010, p. 160).

The Constitution protects both positive and negative freedoms concerning religion; positive in reference to the freedom to openly hold beliefs and practice through worship and teaching; negative in terms of not being forced into practice or acting in opposition to one's personal creed (Jukier & Woehrling, 2010, pp. 160-161). In this respect, the fundamental freedoms likewise protect those who practice atheism, agnosticism, or indifference. Within the judicial branch, freedom of religion is accommodated under the guise of neutrality, meaning that a plaintiff claiming adherence to a religious principle must first prove its existence, then showcase sincerity in belief (Jukier & Woehrling, 2010, pp. 161-62).

---

[5] *The Constitution Act, 1982, provided Canada with sole jurisdiction concerning changes to the Constitution without British involvement (Azzi, 2012).*

Contemporaneous America and its Fundamental Ideology

Unlike Canada, demarcated as the secular component, the United States is significantly more religious. The reasoning for this disparity is resultant of sectarian dissent; while the United States is primarily composed of rival Protestant groups such as Methodists or Baptists, the Canadians adhered to churches with a history of state support (Lipset, 1989, pp. 74-75). It is this differentiation that Canada lacks, because no settler brought the accultured dissent avowed via denominational conflict to Canadian soil (Lipset, 1989, p. 75).

Whereas Canadians valued community and hierarchy, the American spirit materialized from a competitive and individualistic culture borne by capitalistic economies. In Lipset's words, "Since most of these sects were congregational, not hierarchical, they fostered egalitarian and populist values that were anti-elitist. Hence, both the political and religious ethoses reinforced each other" (Lipset, 1989, p. 75). America has a system of 'voluntary religion,' wherein religious groups are not supported by the state; 'voluntarism' is the phenomenon that serves as the explanation for the strength of organized religion in the United States and the implementation of religious foundations in contemporaneous American institutions (Lipset, 1989, p. 75). As noted by Samuel Huntington, Americans provide their nation with a role which mimics the functions of a church (Lipset, 1989, p. 77). In other Christian nation-states, state churches have called for deference and obedience under the sovereign authority, however, American sectarians held that an individual should abide by their conscience rather than adjourn to the state (Lipset, 1989, p. 77). For example, Americans grew increasingly wary of deliberately entering into war despite their long history of expansionism and colonialism, attributed to the strength and growth of religious fervor. To elucidate via historical precedence, "The passion unleashed by the anti-Vietnam War movement was strongly related to basic aspects of the religiously derived American value system" (Lipset, 1989, p. 78). Similar animosity was noted in terms of American entry into the Second World War; had the U.S entered for any reason other than as a defensive[6], it is clear that a large proportion of civil society would

---

[6] The United States formally entered World War Two resultant of an attack on Pearl Harbour by the Japanese.

have blatantly opposed the decision even after Congressional consent (Lipset, 1989, p. 78).

In comparison, Canadians, who were dominated by churches, did not emphasize a likewise skew towards moralism and manifest destiny (Lipset, 1989, p. 79). Canada has been molded by a steady and relatively non-violent departure from Great Britain. Rather than assimilate distinct groups into one, it has diverged from expression of one 'national truth' as evidence for it's conviction and faith (Lipset, 1989, p. 79). In contrast, Americans "...made their dramatic break with England and set out to establish a distinctive nation. Their "charter myth" included the belief that the country had been founded by God to give leadership to the world" (Lipset, 1989, p. 79).

The innate philosophies, historical dogmas, and organizing principles are intrinsic to their foundational orientation contemporaneously. America is the unsullied bourgeois fragment, liberalism untouched by tory deviations, unlike English Canada which is heterogenous (as it carries philosophies of both liberal and conservative beliefs due to emigration) (Horowitz, 1966, p. 148). For these reasons is Canada the secular state carefully manning their extension of toleration towards religion, whereas America remains curtailed by its emphasis on religious 'oneness'.

Bibliography

Azzi, S. (2012, February 6). Constitution Act, 1982. The Canadian Encyclopedia. https://www.thecanadianencyclopedia.ca/en/article/constitution-act-1982.

Baracskay, D. (2009). https://www.mtsu.edu/first-amendment/article/1372/puritans. THE FIRST AMENDMENT ENCYCLOPEDIA. https://www.mtsu.edu/first-amendment/article/1372/puritans.

Behiels, M. D. (1999, July 26). The "Quiet Revolution" to the present. Encyclopaedia Britannica. https://www.britannica.com/place/Quebec-province/The-Quiet-Revolution-to-the-present.

Bowal, P. (2012, November 1). Whatever Happened To... Roncarelli v. Duplessis. LawNow. https://www.lawnow.org/whatever-happened-to-roncarelli-v-duplessis/.

CANADIAN LAW AND RELIGION: THE QUIET REVOLUTION. University of Toronto Libraries. (n.d.). https://exhibits. library.utoronto.ca/exhibits/show/canadianlawandidentity/ cdnlawreligion/cdnlawreligionrevolution.

Freedom of Worship Act. Les Publications du Québec. (n.d.). http:// legisquebec.gouv.qc.ca/en/ShowDoc/cs/L-2.

Hall, A. J. (2006, February 7). Royal Proclamation of 1763. The Canadian Encyclopedia. https://www.thecanadianencyclopedia. ca/en/article/royal-proclamation-of-1763.

Horowitz, G. (1966). Conservatism, Liberalism, and Socialism in Canada: An Interpretation. The Canadian Journal of Economics and Political Science, 32(2), 143–171.

An Institutional History of Religious Freedom in Canada. Cardus. (2020, April 3). https://www.cardus.ca/research/law/reports/an-institutional-history-of-religious-freedom-in-canada/.

Jukier, R., & Woehrling, J. (2010). Religion and the Secular State in Canada. Religion and the Secular State: National Reports, 155–191. https://ssrn.com/abstract=2002271.

Lipset, S. M. (2007). In search of Canadian Political Culture. Continental Divide: The Values of the United States and Canada . https://books-scholarsportal-info.myaccess.library.utoronto.ca/ uri/ebooks/ebooks3/upress/2013-08-25/1/9780774855990.

McConnell, W. H. (2006, February 6). Constitution Act, 1867. The Canadian Encyclopedia. https://www.thecanadianencyclopedia. ca/en/article/constitution-act-1867.

Nesbitt-Larking, P., Adams, M., Charlton, M. W., & Barker, P. (2012). Is the Canadian Political Culture Becoming Americanized? In Crosscurrents: Contemporary Political Issues (7th ed., pp. 2–34). Nelson College Indigenous.

Paradis, P., & Karbani, T. (2017, May 9). The Significance of the Charter in Canadian Legal History. LawNow. https://www.lawnow.org/significance-charter-canadian-legal-history/#:~:text=The%20patriation%20battle%20between%20Prime,it%20on%20April%2017%2C%201982.

The Quebec Act, 1774 (Plain-Language Summary). The Canadian Encyclopedia. (2020, January 17). https://www.thecanadianencyclopedia.ca/en/article/the-quebec-act-1774-plain-language-summary.

Roncarelli v. Duplessis, SUPREME COURT OF CANADA (SCR January 27, 1959). https://scc-csc.lexum.com/scc-csc/scc-csc/en/item/2751/index.do.

Wiseman, N. (2007). In Search of Canadian Political Culture. UBC Press. UBC Press. https://www.ubcpress.ca/asset/9386/1/9780774813884.pdf.

Targeting Recidivism. Are day-fines an equitable replacement for tariff-based penalties?

Minahil Syed, Peter Anto Johnson, John Christy Johnson

A **standardized** fine determines the severity and cost of the crime, not of the criminal. While this does ensure impartiality before the law, does it sufficiently affect rates of recidivism and compliance?

Increasing the penalty associated with fines in proportion to an individual's income serves as a deterrent for those who can intentionally choose to ignore the rate in favour of re-offending; it is a matter of cost-benefit analysis. The current structure should be focused on targeting defiance and uprooting recidivism, the decision to feign ignorance has bred anomie towards public security and wellbeing. For instance, individuals with a higher disposable income may feel comfortable parking in a disabled spot when they themselves are not disabled – simply because it is more convenient, even if they do incur a 'small' fine of 70$. Tariff-fines, set at a fixed amount relative to the incurred offense, are the predominant form of economic sanction used in both the United States and Canada; their nature makes them entirely regressive and biased (Colgan, 2017, p. 55). Fines presently target the impoverished individual's socioeconomic status rather than serve as a penalty for a crime, hence, the question arises, are fines a mechanism to target reoccurrence or simply to unjustly punish the poor?

Paying fines is intended to serve as a deterrent, and while they should not factor in an individual's market value (income/wage/wealth) when establishing the penalty, a fine incurred resultant of breaking a law should serve as an equitable punishment for all. Fines imposed without a regard for income entrap low-risk offenders in a cycle of debt (or jail, should the fines exceed their wage) while letting high-income offenders run free without significant financial consequence (Schierenbeck, 2018, p. 1870). Fixed fines, while equitable in their intention, fall flat in encouraging respect for the law. It should not be more burdensome for a poorer individual than a wealthier one, no one should be able to 'buy' themselves out from of an offense. Legal discretion is avowed during trials, hence fines are not necessarily flat, and while a judge has the option to reduce fines as deemed necessary, they cannot directly **increase**[7] the charge or fine (Tremblay, 2010). While the ability to lower damages is befitting for the poor, there is no prospect to markedly inconvenience, or hold liable, the rich.

Relying on a **singular** metric in order to decide the judgement of a fine has the ability to introduce systemic or personal bias into issuance (Popping, 2012, pp. 7-9). An equitable measure would not one-sidedly provide a ruling via judgement of an individual's income. Nonetheless, when considering penalty structures, it should be emphasized that a crime should not serve as a net benefit for law enforcement or local government to profiteer off of, in other words, crime should not be incentivized.

The "day-fine" model gages penalties in proportion to the income of the offender and the seriousness of the crime (based on a

---

[7] *The right to due process ensures that all individuals are bestowed with an unbiased and fair application of the law before imprisonment, seizure of property, or their verdict ("Due process"). Due process thus requires notice of the charges being held against the defendant and the opportunity to dispute against them. Hence, a perpetrator cannot be convicted of a more serious offence than with which they were charged with resultant of the double jeopardy clause ("Section 11(h) – protection against double jeopardy").*

relative scale) (Colgan, 2017, pp. 56-57). The model would assign a numerical value, 'units,' to the crime, which is then proportional to the amount of days that an individual would work in order to pay off the fee. The final amount would be calculated by multiplying days worked and individual income, theoretically creating fines which would suffice as an equal fiscal burden for anyone based on their socioeconomic status (Walsh, 2019). The same crime could accost a wealthy individual $6,000 for speeding, while the average earner would be charged $60. In Finland, income evaluation is predominantly done via self-assessment, which facilitates an incentive to lie, nonetheless, as courts have access to tax data, lying is relatively uncommon (Schierenbeck, 2018, p. 1894).

In August 1988, the Criminal Court of Richmond County underwent means-tested day-fine experiments in Staten Island - which accounted as one of the five boroughs under their jurisdiction within New York. Economic sanctions of this sort are borrowed concepts implemented initially in Scandinavia (1920s-30s) and then West Germany (1960s-70s) as substitutes to short-term incarceration (Greene, 1990, p. x). Although tariff-fines are regarded as unfairly punitive for those who are financially unstable, the concept arose as a corrective to the tough-on-crime legislation (rapid increase in mass incarceration and probation) that developed in the 1980s within the United States (Colgan, 2017, p. 55). During this period, economic sanctions gained momentum as an equitable mechanism to combat civil offenses while taking the incarceration versus rehabilitation line of reasoning into account (Colgan, 2017, pp. 55-56). The day-fine system addresses the predicament between the monetization of criminality and burden of accountability for offenders, because,

"When fined, the offender quite literally is made to pay his or her debt to society. When the fine can be flexibly adjusted to fit both the gravity of the offense and the offender's means (as day fines permit), it does not destroy the offender's ties to family and community. It can also be an important source of revenue and does not require the resources of additional administrative agencies for implementation" (Greene, 1990, p. 3).

Graduated sanctions likewise respond to revelations of precincts overcharging offenders knowing they were unable to pay and jailing them for failure to do. For instance, in 2015, the city of Ferguson claimed a lawsuit which attested that the municipality had devised a scheme to yield profit via uninhibited targeting; the charge indicates that, "...city officials violated the US constitution by jailing people without making a meaningful inquiry into their ability to pay court-ordered fines or offering them a lawyer" (Gambino, 2015). In 2013, the (estimated) net population of Ferguson amounted to 21,111 people, and in the same year, courts issued 32,975 warrants for arrest, the majority for driving violations (Gambino, 2015).

Reformation via graduated sanctions serve as equitable mechanisms to transform a judicial and executive structure plagued with inefficiency, recidivism, and systemic bias, while concurrently meeting revenue-generation goals via fair sentencing (Colgan, 2017, pp. 53-55).

Bibliography

Colgan, B. A. (2017). Graduating Economic Sanctions According to Ability. Iowa Law Review, 103, 53–112. https://ilr.law.uiowa.edu/print/volume-103-issue-1/graduating-economic-sanctions-according-to-ability-to-pay/.

The Editors of Encyclopaedia Britannica. (n.d.). Due process. Encyclopaedia Britannica. https://www.britannica.com/topic/due-process.

Gambino, L. (2015, February 9). Ferguson lawsuit sues city over alleged jailing of people too poor to pay fines. The Guardian. https://www.theguardian.com/us-news/2015/feb/09/ferguson-lawsuit-sues-city-jailing-poor-pay-fines.

Greene, J. (1990, July). THE STATEN ISLAND DAY FINE EXPERIMENT. Vera Institute of Justice. https://www.vera.org/publications/the-staten-island-day-fines-experiment.

Popping, H. (2012). Do Higher Fines Reduce Recidivism? Evidence from a Twenty Percent Increase in (Dutch) Traffic Fines (thesis). Semantic Scholar. Retrieved from https://api.semanticscholar.org/CorpusID:155129102

Schierenbeck, A. (2018). The Constitutionality of Income-Based Fines. The University of Chicago Law Review, 85(8), 1869–1925. https://lawreview.uchicago.edu/publication/constitutionality-income-based-fines#:~:text=Income%2Dbased%20fines%20could%20help,proceedings%2C%20and%20periods%20of%20incarceration.

Section 11(h) – protection against double jeopardy. Department of Justice. (n.d.). https://www.justice.gc.ca/eng/csj-sjc/rfc-dlc/ccrf-ccdl/check/art11h.html.

Tremblay, L. (2010, April 16). From substantive due process to substantive principles of fundamental justice. Allard Research Commons. https://commons.allard.ubc.ca/theses/317/.

Walsh, M. (2019, November 21). Day Fines: A First Step in Ending Mass Incarceration. Brown Political Review. https://brownpoliticalreview.org/2019/11/day-fines-a-first-step-in-ending-mass-incarceration/.

Where have the people gone?

Minahil Syed, Peter Anto Johnson, John Christy Johnson

The "End of Public Space" thesis (Don Mitchell) is the idea that publicness has to be constructed. Mitchell states that **truly** public spaces are now over because they have been privatized; redevelopment of private spaces now revolves around consumption activities rather than sociable uses (Mitchell, 1995, p. 115).

As the world around us industrializes more, we begin to note that public activities have become commercialized, demand a fee to participate in, or are in a state of eminent domain. Are we participating in a tragedy of the commons guided by the invisible hand - wherein the commons are depleting due to a corporate monopoly of previously public land? Mitchell addresses this issue via criticism of the University of California (UC) People's Park in the 1990s; as police and bulldozers cleared the soil and grassy hills, protestors firmly contested the "...expansion of corporate control over the fabric of cities" (1995, p. 109). The goal of the university was to create a 'controlled environment in a highly urbanized area', and in aspiring for so, the central belt where homeless people slept and political organization was coordinated, would be converted into a volleyball court with walkable pathways (pp. 109-110). Likely to drive away from areas for homeless people to unobtrusively settle, although such a statement was repudiated by UC spokesperson Jesus Mena who infamously declared, "We have no intention to kick out the homeless. They will still be there when the park changes, but without the criminal element that gravitates toward the park" (Mitchell, 1995, p. 110). For homeless people in the San Francisco Bay Area, People's Park was one of the few public venues wherein they could live without intervention. Activists felt that the renovation of the space inherently dimmed the park's figurative role as a refuge and counterculture. In their eyes, its characterization as an 'unrestricted retreat' marked by the absence of corporate institutions was appropriated (Mitchell, 1995, pp. 110-113). Public spaces are the only venues where individuals can represent a genuine component of what constructs 'the public,' insofar as they remain invisible, they "...fail to be counted as legitimate members of the polity" (Mitchell, 1995, p. 115).

When public spaces cease to exist, what does this mean for the individual? Public space must be recognized as an area *of* and *for* the commons. The industrialization has created a narrow judgement of what growth and progress encompass, making it seem as though there is only one teleological path to urban development – meaning, high-density housing, narrow pathways, lack of open space for physical activities (e.g. bike-riding or skateboarding), and anti-homeless architecture. The public realm is significant not only for its role in socialization but as a precondition for a formally functioning society (Bibeva, 2012, p. 9). As Madanipour elucidates, activities in our contemporaneous age require little to no social contact, resultant of technologies which have made it possible to divulge in external services (i.e. exchange of information, money transfer, shopping) without leaving the home (innocuously deemed the private realm). This disposition, "...inevitably influences the process of shaping and expressing social relationships...social life is beginning to lack spatial manifestation. As a result, the public realm can be found in the cyber world as much as in the physical world" (Bibeva, 2012, p. 9). In this domain, effectiveness and rationality should then be the guiding principles in the construction of the public realm (Bibeva, 2012, p. 9). There is an association between the quality of public spaces and the sense of community fostered (Bibeva, 2012, p. 14).

Well-designed areas naturally bring the commune together while simultaneously attracting investment; they are responsible for identity-shaping and overcoming social exclusion via discernment of how socio-spatial segregation occurs (a result of infrastructure that is unable to discern between community needs and what modernist planning demands) (Bibeva, 2012, p. 14).

Socialization in our present culture has changed, albeit, whether public space has ended is another question. Nonetheless, it is contingent on the recognition that the commons have been marketized and privatized; as stated by Gehl, "It is of prime importance to recognize that it is not buildings, but people and events, that need to be assembled" (2011, p. 81). It is this philosophy that should guide the emergence beyond modernist architecture and into an age of neotraditional urbanism wherein expansionism is based on the interconnectedness of public spaces and the creation of linkages within the community (Bibeva, 2012, p. 15).

Bibliography

Bibeva, I. (2012). *Public Space and its Role for Segregation, Identity and Everyday life. A Case study of Östbergahöjden and its Square (thesis)*. Stockholm.

Gehl, J. (2011). *Life Between Buildings: Using Public Space. Ebook Central Perpetual and DDA* (6th ed.). Washington DC: Island Press. https://librarysearch.library.utoronto.ca/permalink/01UTORONTO_INST/fedca1/cdi_proquest_ebookcentral_EBC3317590.

Mitchell, D. (1995). The End of Public Space? People's Park, Definitions of the Public, and Democracy. *Annals of the Association of American Geographers*, 85(1), 108–133. http://links.jstor.org/sici?sici=0004-5608%28199503%2985%3A1%3C108%3ATEOPSP%3E2.0.CO%3B2-M.